Writing API Tests with Karate

Enhance your API testing for improved security and performance

Benjamin Bischoff

BIRMINGHAM—MUMBAI

Writing API Tests with Karate

Group Product Manager: Gebin George

Publishing Product Manager: Kunal Sawant

Senior Editor: Rounak Kulkarni

Technical Editor: Pradeep Sahu

Copy Editor: Safis Editing

Project Coordinator: Deeksha Thakkar

Proofreader: Safis Editing

Indexer: Manju Arasan

Production Designer: Shyam Sundar Korumilli

Developer Relations Marketing Executives: Sonia Chauhan and Rayyan Khan

Business Development Executive: Kriti Sharma

Production reference: 0130423

Published by Packt Publishing Ltd.
Livery Place
35 Livery Street
Birmingham
B3 2PB, UK.

ISBN 978-1-83763-826-0

www.packtpub.com

I dedicate this book to my wife, Anne, and my daughter, Frida, who gave me so much love, time, and support to finish this project without going crazy. I love you! Thanks a million!

– Benjamin Bischoff

Foreword

It gives me great pleasure to write this foreword for the first-ever book dedicated to Karate.

My journey with Karate began with a simple desire to help teams struggling with test automation. I have always been driven by the belief that there had to be a simpler, better, and more maintainable way to automate tests. As the creator of Karate, I am indeed thrilled to see how far it has come, and to witness the impact it has had on teams and enterprises all around the world. I am grateful to the community that has embraced and contributed to Karate's growth, making it a powerful tool for organizations.

What started out as a small side-project focused on API testing is now an open-source company, *Karate Labs*. Since the first version of Karate was released six years ago, we are trusted by 1000's of organisations and are aware we have so much more to do as we scale Karate Labs as a responsible open-source company.

Hand-on-heart, I have always wondered whether I should write a book on Karate. So, when I came to know that Benjamin is writing a book on Karate, I felt a sense of relief. There's nothing better than having an accomplished thought leader write a book on Karate. I deeply respect Benjamin's work in the test-automation and open-source domain and had the pleasure of hosting him on one of our customer webinars. You can count on him to give you an informed, balanced perspective - and with great attention to detail.

Thank you, Benjamin - for your dedication and expertise in crafting this guide to Karate. I am confident that this book will be an invaluable resource for both beginners and seasoned professionals, enabling them to harness the full potential of Karate and take their test automation efforts to new heights.

Sincerely,

Peter Thomas

Co-founder of Karate Labs and creator of the Karate test-automation framework

Contributors

About the author

Benjamin Bischoff decided in 2017 to make test automation his main career after being a full-time developer for 15 years. He works as a test automation engineer in Trivago's product foundation backend QA team, where his focus lies on backend, API, and database automation, as well as build and test pipelines. Prior to that, he took care of frontend automation and the development and maintenance of Trivago's in-house end-to-end test framework. Benjamin is the author and maintainer of two open source projects for Cucumber BDD parallel test execution and reporting. He is also a conference speaker and writes about testing and automation topics on his website, `https://www.softwaretester.blog`.

I would like to thank the people who were close to me and gave me feedback. My brother, my parents, and the entire Trivago QA team, especially Saurav Sharma for his technical contribution and Karan Tohan for convincing me about Karate. I also thank Peter Thomas and Kapil Bakshi, for all their work and dedication. Thank you, Packt team, for all the help!

About the reviewer

Saurav Sharma is a seasoned software quality assurance professional with over 12 years of experience in the field. Currently, he works as a QA lead for Trivago at their headquarters in Düsseldorf, Germany, where he collaborates with an excellent team of QA professionals, focusing on backend services. Originally, he is from Katras, a small town in the Dhanbad district of Jharkhand, India, and holds a BTech degree in information technology. He enjoys technical and people leadership and has worked in various capacities with web and mobile applications for companies including Oracle, Rocket Software, Allscripts, and Tech Mahindra. He strongly believes that test automation and CI are vital for ensuring speed and quality in software development.

Table of Contents

3

Writing Basic Karate Tests   55

4

Running Karate Tests   89

5

Reporting and Logging   117

Part 2: Advanced Karate Functionalities

6

More Advanced Karate Features   139

7

Customizing and Optimizing Karate Tests   177

8

Karate in Docker and CI/CD pipelines   209

9

Karate UI for Browser Testing   237

10

Performance Testing with Karate Gatling   269

Preface

The Karate framework is a relatively new software testing solution that aims to bridge the gap between exploratory testing and code-heavy automation. It has a strong focus on being easy to learn but still providing complete functionality for API testing and beyond.

This book first explains the core concepts and ideas of the framework, how to set up new test projects, and how to write effective API tests. It further covers the various ways to run these tests in different settings and environments and explains various reporting and logging options, including completely custom ones. It also covers advanced Karate features, such as extending the functionality of the framework with custom JavaScript and Java code and even using it in Docker and **continuous integration/continuous delivery** (**CI/CD**) pipelines.

In addition, the book covers two of Karate's lesser-known capabilities: Karate UI for browser testing and performance testing with Karate Gatling. This comprehensive guide is a useful resource for anyone looking to learn about the Karate framework and implement it in their software testing practice.

Who this book is for

If you are a QA engineer or developer who is already familiar with APIs and wants to cover them through automated testing, this book is for you. Even if you have done other forms of testing and want to know more about API testing, this book can help you understand the core ideas and approaches of these tests. And even if you already use other frameworks and techniques for API testing, you will learn here what sets Karate apart from its competitors. To that end, you'll find a lot of information in this book to help you decide whether Karate can be a good replacement for your current toolset.

What this book covers

Chapter 1, *Introducing Karate's Core Concepts*, introduces the Karate testing framework, which provides an overview of its strengths, key concepts, and special features.

Chapter 2, *Setting Up Your Karate Project*, provides a step-by-step guide for setting up a new Karate test project, including **integrated development environment** (**IDE**) preparation, dependency management with Maven, and basic configuration using the Karate configuration file.

Chapter 3, *Writing Basic Karate Tests*, introduces writing the first API tests using Karate, which examine return codes and responses while ensuring code efficiency through Karate and Gherkin mechanisms.

Chapter 4, *Running Karate Tests*, explores various methods for triggering and running Karate tests, including considerations for running tests in CI/CD pipelines, selecting test cases, and executing tests in parallel to optimize runtime.

Chapter 5, Reporting and Logging, gives an overview of Karate's built-in reporting and logging capabilities for effective troubleshooting, plus a guide for integrating third-party reporting solutions.

Chapter 6, More Advanced Karate Features, explores concepts and techniques for creating more complex test cases, including setting and checking headers, cookies, and authentication tokens, testing GraphQL APIs, and switching between different test environments.

Chapter 7, Customizing and Optimizing Karate Tests, shows how to create custom functionality in Karate through JavaScript functions and Java interoperability, as well as how to write custom Karate hooks to react to Karate events.

Chapter 8, Karate in Docker and CI/CD Pipelines, integrates Karate tests into CI/CD pipelines to establish a fully automated test setup using the example of GitHub workflows.

Chapter 9, Karate UI for Browser Testing, introduces Karate UI, a special module for browser-based test automation, and explores how this alternative approach to **user interface** (**UI**) testing fits into the Karate ecosystem.

Chapter 10, Performance Testing with Karate Gatling, explores the integration between Karate and the popular Gatling framework for load and performance testing, which reuses existing Karate scenarios.

To get the most out of this book

Basic Java and JavaScript knowledge, as well as a basic understanding of the testing methodology, will help you to find your way around faster. However, in this book, I also try to cover the required aspects in a way that is understandable even without this knowledge.

Software covered in the book	Operating system requirements
Java SDK	Windows, macOS, or Linux
Maven	Windows, macOS, or Linux
Karate Standalone	Windows, macOS, or Linux
Google Chrome	Windows, macOS, or Linux
IntelliJ IDEA	Windows, macOS, or Linux
Visual Studio Code	Windows, macOS, or Linux
Postman	Windows, macOS, or Linux
MySQL	Windows, macOS, or Linux
phpMyAdmin	Windows, macOS, or Linux
Git	Windows, macOS, or Linux
Docker	Windows, macOS, or Linux

You will not need any commercial software or tools to follow along. I deliberately wanted to use free and open source software so everyone can use it without spending additional money.

If you are using the digital version of this book, we advise you to type the code yourself or access the code from the book's GitHub repository (a link is available in the next section). Doing so will help you avoid any potential errors related to the copying and pasting of code.

Download the example code files

You can download the example code files for this book from GitHub at `https://github.com/PacktPublishing/Writing-API-Tests-with-Karate`. If there's an update to the code, it will be updated in the GitHub repository.

We also have other code bundles from our rich catalog of books and videos available at `https://github.com/PacktPublishing/`. Check them out!

Download the color images

We also provide a PDF file that has color images of the screenshots and diagrams used in this book. You can download it here: `https://packt.link/IIe3X`.

Conventions used

There are a number of text conventions used throughout this book.

`Code in text`: Indicates code words in text, database table names, folder names, filenames, file extensions, pathnames, dummy URLs, user input, and Twitter handles. Here is an example: "Gherkin files have the `.feature` extension and always start with a `Feature:` line followed by a description."

A block of code is set as follows:

```
Package blog.softwaretester.gherkin;
import io.cucumber.java.en.Given;
import io.cucumber.java.en.Then;
import io.cucumber.java.en.When;
```

When we wish to draw your attention to a particular part of a code block, the relevant lines or items are set in bold:

```
public class StepDefinitions {
    @Given("I am on the Web shop homepage")
    public void goToHomepage() {
        System.out.println("Go to homepage");
    }
```

Any command-line input or output is written as follows:

```
java -cp C:\Users\bbischoff\Desktop\karate-1.2.1.RC1\karate.jar
com.intuit.karate.Main
```

Bold: Indicates a new term, an important word, or words that you see onscreen. For instance, words in menus or dialog boxes appear in **bold**. Here is an example: "To trigger one of the saved requests, click **Send**."

> **Tips or important notes**
> Appear like this.

Get in touch

Feedback from our readers is always welcome.

General feedback: If you have questions about any aspect of this book, email us at customercare@packtpub.com and mention the book title in the subject of your message.

Errata: Although we have taken every care to ensure the accuracy of our content, mistakes do happen. If you have found a mistake in this book, we would be grateful if you would report this to us. Please visit www.packtpub.com/support/errata and fill in the form.

Piracy: If you come across any illegal copies of our works in any form on the internet, we would be grateful if you would provide us with the location address or website name. Please contact us at copyright@packt.com with a link to the material.

If you are interested in becoming an author: If there is a topic that you have expertise in and you are interested in either writing or contributing to a book, please visit authors.packtpub.com.

Share Your Thoughts

Once you've read *Writing API Tests with Karate*, we'd love to hear your thoughts! Scan the QR code below to go straight to the Amazon review page for this book and share your feedback.

https://packt.link/r/1837638268

Your review is important to us and the tech community and will help us make sure we're delivering excellent quality content.

Download a free PDF copy of this book

Thanks for purchasing this book!

Do you like to read on the go but are unable to carry your print books everywhere? Is your eBook purchase not compatible with the device of your choice?

Don't worry, now with every Packt book you get a DRM-free PDF version of that book at no cost.

Read anywhere, any place, on any device. Search, copy, and paste code from your favorite technical books directly into your application.

The perks don't stop there, you can get exclusive access to discounts, newsletters, and great free content in your inbox daily

Follow these simple steps to get the benefits:

1. Scan the QR code or visit the link below

https://packt.link/free-ebook/9781837638260

2. Submit your proof of purchase
3. That's it! We'll send your free PDF and other benefits to your email directly

Part 1: Karate Basics

In this part, we will introduce the ideas and concepts of Karate and examine what sets it apart from other frameworks. We will set up the system and learn how to create new test projects using Maven. After that, we will start writing our first tests and explore different ways to run them. Finally, we will take a closer look at the reporting and logging options built into Karate and go through how you can integrate your own reports.

This section contains the following chapters:

1

Introducing Karate's Core Concepts

The software testing world is constantly changing. Technologies that were recently state of the art are replaced or turn out to be slow sellers. New frameworks are created all the time to try to solve common testing problems. There are a lot of people who have entrenched opinions about how tests should look, and sometimes this attitude can even block progress. Testing is often still divided into exploratory testing (also called *manual testing*, which I personally dislike) and code-heavy automation solutions (also known as *test automation*), which may require a software expert for development and maintenance but might not be feasible for more hands-on exploratory testing.

In between these extremes, there is plenty of room for innovation. There has been quite some progress here during the last few years, and these new technological solutions are in high demand.

The Karate framework is a rather new contender in the software testing field that has set out to solve many of the problems raised, especially the steep learning curve of other solutions.

We are going to cover the following topics:

- What makes Karate stand out?
- **Behavior-driven development (BDD)** versus Karate
- Supported data types in Karate
- The JavaScript engine
- Java interoperability

Technical requirements

All code examples for this chapter can be found on GitHub at `https://github.com/PacktPublishing/Writing-API-Tests-with-Karate`.

> **Demo code**
>
> It is not necessary to set up the programming environment at this point. All code that is shown here is just to demonstrate Karate's inner workings and core functionality. In the next chapter, we will deal with this thoroughly!

What makes Karate stand out?

At its core, Karate is a test automation framework that is written in Java but does not necessarily require programming skills when used for basic software testing. It is heavily based on the **Gherkin** syntax, made famous by the **Cucumber BDD** framework. In fact, when it was first introduced to the software testing world, it was designed as an extension of Cucumber but quickly evolved into something of its own. However, some of Cucumber's core characteristics are still evident and can be put to good use. For example, the very similar file structure, syntax, and compatibility with the standard JSON reporting format.

The first version was released on February 08, 2017, by Peter Thomas while he was working as a test engineer at Intuit India (`https://www.intuit.com/in`), a large software company in the fintech industry. His motivation was to come up with an easier-to-use and easier-to-understand API testing solution than the in-house framework his team was using at this time.

In Peter's own words,

> *Karate strives to reduce the entry barrier to writing a test and more importantly - reduces the friction to maintain a test, because of how readable tests become.*

Thomas, P. (2017). Karate: Web-Services Testing Made Simple, `https://medium.com/intuit-engineering/karate-web-services-testing-made-simple-366e8eb5adc0`

Intuit suggested making his work open source in order to test how this new approach would be received by the community.

At the time of writing this book, the Karate framework sports more than 6.6k stars on GitHub (`https://github.com/karatelabs/karate`) and even received a GitHub grant for open source software in India.

Peter has apparently struck a nerve with the testing community!

In 2021, after this successful open source launch and adaption period, Peter decided to found Karate Labs (`https://www.karatelabs.io`), together with Kapil Bakshi, in order to work full-time on the further development and promotion of the framework and to offer some new paid products and services around Karate.

> **No paywall**
>
> The paid services by Karate Labs include subscription-based plugins for IntelliJ IDEA and Visual Studio Code with multiple tiers. In this book, however, we will focus exclusively on the freely available core framework and its various open source components and tools.

Interest in the Karate framework has not waned since then, as can be seen by the plethora of questions on **Stack Overflow** and its GitHub issues page. Since a lot of information is spread among many different online articles, social media channels, blog posts, and podcasts, I decided to write a book about it to provide a structured, comprehensive, and complete introduction.

In the upcoming sections, we will discover how Karate works in more technical and conceptual detail and what distinguishes this solution from other popular testing frameworks that may be more specialized but are also much harder to learn and use.

Also, we will further explore the similarities and differences between the *classic* Gherkin syntax and underlying concepts and Karate's different approach in the next section. We will find out why certain decisions were made by the framework creator and why they were smart moves.

Discovering Karate's strong points

Karate's strengths lie in various areas of test automation, which we will discuss in this section. Later on in the book, we will shine a light on these different areas with a strong focus on the framework's core purpose – testing APIs effectively.

Since there are multiple modules of Karate with different goals, let's quickly walk through them to see how extensive the framework and its components have become over the years.

Testing APIs

Providing a basis for API tests that is easy to read and follow is the main capacity of Karate – it supports the **REST**, **SOAP**, and **GraphQL** APIs and can be extended to handle other types of APIs and data formats that are not as common. All other framework modules that cover different kinds of testing were gradually added and implemented with the same basic mechanisms and concepts. By focusing on its straightforward domain-specific Gherkin-based language, Karate can map a lot of API test functionality without having to write elaborate code for communicating with web services or handling various payloads and data exchange formats.

Additionally, Karate includes many powerful assertions to verify response data, and its structure, especially **JSON** and **XML**, are handled exceptionally well. This makes it a perfect fit for a lot of modern APIs that use these formats.

We will explore all of this in *Chapter 3, Writing Karate Tests*.

> **Fun fact**
> Even though Karate is often associated with JSON, XML support was implemented first in its initial prototype before JSON handling was added!

Test doubles

Testing microservice infrastructures or web services calling other services or databases can get very complex and challenging. Typically, the approach is to spin up all components that make up a full system in order to test it thoroughly. This makes it hard to run tests for a single component because each test needs to go through multiple layers. This means that other dependent services need to be very predictable and reliable in such a setup so you can obtain meaningful and non-flaky test results.

Test doubles, a term coined by agile software development consultant Gerard Meszaros (`http://xunitpatterns.com/gerardmeszaros.html`), solve this problem. It groups together all forms of fake objects (or fake APIs) that can be used instead of real objects in tests. They behave as if you were making real requests and data transfers but deliver the same predictable responses every time. This allows you to perform testing on a small part of the system without having to directly use all real dependencies and services. Thus, the tests can run faster and more securely because you are not affected by network problems, bad connections, database inconsistencies, and other failures.

Karate can define such doubles easily using nothing but Gherkin, as we will see in *Chapter 6, More Advanced Karate Features*!

Performance tests

Karate's main strength in the performance testing field is that it can reuse existing scenarios. The same test cases that are used as acceptance tests can seamlessly act as **Gatling** performance tests. This means that it is possible to simulate multiple clients or connections that interact with your APIs in order to check their resilience and how they hold up under heavy loads. By integrating the popular Gatling framework, Karate gains a huge advantage here because it frees us from using completely separated solutions, third-party frameworks, and libraries that would require additional research, learning, and onboarding.

Also, due to its ability to reuse already implemented test scenarios, we do not need to spend time and resources duplicating them in another programming language just for the sake of performance tests.

In *Chapter 10, Performance Testing with Karate Gatling*, we will look deeper at this important testing topic.

Browser automation

The automation of websites and browser-based user interfaces has become much more important in recent years. More and more user-facing software is available as web applications, which can be very complex in design. Also, a lot of functionality and logic can be wired into the frontend, making it necessary to thoroughly test this area as well.

Karate UI is special here because it combines the **Chrome DevTools Protocol** (**CDP**) and the **WebDriver** protocol, so it can accomplish pretty much everything that other popular tools such as **Selenium**, **Cypress**, and **Playwright** can do.

As we will see in *Chapter 9, Karate UI for Browser Testing*, this module was one of the last ones added to the Karate framework to cover this testing field as well. Among other things, it provides full debugging capabilities and even allows freely mixing API and UI automation to handle even more complex use cases.

Desktop automation

Karate also supports automating desktop applications through its integration with the **Robot** framework. This is an entirely different use case than the other ones that are discussed in this book, but maybe this will contribute to your decision to give Karate a try.

If you want to know more about this part of Karate, check the documentation at `https://github.com/karatelabs/karate/tree/master/karate-robot`.

Now that we have gotten to know Karate's core uses, let's find out about its main features next.

Core features

Since many useful and often needed functions of other specialized test frameworks are already implemented in the core framework, you will reach useful results fast and without a steep learning curve. This makes Karate ideal for quickly implementing a large number of test cases and working off low-hanging fruits (things that take only a little time while having a great benefit).

Karate's big advantage is the consistent syntax across test types. This makes switching between different test projects and contexts easier and faster, as there is no need for extensive training. Also, it keeps the test suites very clean, easy to understand, and maintainable when written well.

If, however, you require expanded functionality, you are free to add it as either Java or JavaScript code, depending on which language features can solve your problem better and your respective skills with each of them.

Let's look at some of its most useful aspects!

Parallel test execution

Often, extensive test suites contain a lot of different scenarios that differ greatly in individual runtime. Executing these tests in sequence, meaning one after the other, can lead to an unnecessarily long feedback time. Many other test frameworks have parallelization built in, at best, as an afterthought or require additional tools for this. Also, it can be complex to execute and evaluate (for example, because test reporting cannot deal with parallel scenarios correctly). Karate has this approach on board as a basic feature for all different test types.

As we will later see in *Chapter 4, Running Karate Tests*, we will even be able to visualize and track how tests are run on different threads, making this a powerful option for more efficient test runs.

Data-driven testing

In *Chapter 3, Writing Basic Karate Tests*, we will discover that the Karate framework even supports data-driven tests. That means that one or more test scenarios can be reused with multiple sets of data instead of writing separate ones for each individual use case. This reduces code duplication and errors caused by having to change multiple code locations across a lot of feature files if requirements change.

The unique aspect of Karate in this regard is that these records of data can even be defined in **JavaScript Object Notation (JSON)** or **comma-separated values (CSV)** formats. Also, they are not required to be static data as in a Cucumber-based framework but can be generated dynamically.

Auto-generated test reports and logs

Karate has its own built-in reporting library that generates informative HTML-based test reports that show what features and scenarios were run along with requests, responses, and additional information such as tags, screenshots, and even a timeline of parallel runs. These help greatly in analyzing test runs and can act as a form of documentation as well. In case you need to display custom information, Karate even includes a full templating engine that can be programmed to display any additional data you may need.

Additionally, all Karate actions, up to every single request and response, are logged. Even with the most beautiful test reports, it is often simple text logs that provide the most information about errors that have occurred.

We will encounter both variants of debugging, including custom configuration and extension, in *Chapter 5, Reporting and Logging*.

First-class support of common data formats

Karate, through its Java and JavaScript roots, can handle pretty much any data format found on the web today. This does not only mean processing them but also making assertions, performing schema validations, and efficient fuzzing matching of results. Later in this chapter, we'll take a closer look at the formats that are natively supported. Additionally, we will explore this topic further in *Chapter 3, Writing Basic Karate Tests*, and *Chapter 4, More Advanced Karate Features*.

Beginner friendliness

One of the best characteristics of the Karate framework is that it allows users to start using it as a solution where no coding is necessary and gradually realize more complex test cases by writing Java or JavaScript code on top of it. The Karate **domain-specific language (DSL)** will be discussed in more detail a little later in this chapter.

When using Karate, it is worth mentioning that it is usually just one single dependency you need that covers most functionality (for example, API testing, UI testing, and performance tests), making this rather simple to set up.

Additionally, there is even a standalone executable version of Karate that can do everything that the Java library can do but without requiring experience with setting up a Java project. This is useful for quickly trying out Karate without extensive preparation. For a production environment, I would nonetheless recommend the Maven setup that we will see in *Chapter 2, Setting Up Your KarateProject,* and *Chapter 4, Running Karate Tests*.

Built-in support of different environments

If you have a mature test suite, the goal is typically to run it within a continuous delivery pipeline on different development environments, for example staging, user acceptance, and live. Often, these environments have different URLs, endpoints, access rules, and configurations. So, it is very convenient if the test framework offers a simple way to keep these configurations inside the test repository and switch easily between them.

Karate offers this as a native functionality as we will see in *Chapter 6, More Advanced Karate Features*.

Read-eval-print loop

A **read-eval-print-loop** (**REPL**) is a programming environment that reads user inputs, then processes and executes them as code snippets and prints the results. Karate has a command-line REPL, which allows test code to be inserted or replaced at runtime, making it easier and faster to develop tests. This can save time that would otherwise be spent modifying, adapting, and executing a complete test scenario repeatedly. The REPL is a main component of Karate's Visual Studio Code plugin and can also be used for UI tests!

> **One framework for all**
>
> All previously mentioned features are anything but self-evident, and it is an accomplishment not to be undersold that they come together here in one integrated software testing solution!

In this section, we looked at Karate's main use cases and features that make it stand out from other frameworks. Next, we will turn to an aspect of the framework that is controversial to some testers – the difference between the BDD approach and how Karate takes this idea and changes it.

BDD versus Karate

Now that we know about the core ideas and features of the Karate framework, let's look at how it achieves writing meaningful test scenarios without extensive coding.

What is BDD?

BDD comes from the world of agile software development. It focuses on bringing together the different roles in a software project (for example developers, testers, and project owners) and talking about requirements. These requirements should then be put into clear scenarios so that, later, all roles can take these scenarios as a basis for their work.

A core part of BDD is a ubiquitous and simple DSL that can express desired behavior in natural language and is understood by every team member. Its purpose is to form a basis for the development of new features but also be a clear guideline for testing and acceptance. In the following sections, we will look at one of the most well-known and widely used DSLs – Gherkin.

The Gherkin language

If you have been in touch with the Cucumber Open BDD testing tool (https://cucumber.io/), you will have stumbled across the Gherkin language. Gherkin is a very concise and straightforward way to turn acceptance criteria into understandable and well-structured test scenarios. At its core, it is a plain-text format to describe how a system under test should behave. Cucumber can parse this format and link it to custom test code, so it provides a readable wrapper around the technical implementation of tests. This way, other development and QA team members, as well as people with non-technical roles, can read these specifications and know exactly what is expected from the system.

The following is an example of the contents of a typical Gherkin feature file (named webshop.feature):

```
Feature: Web shop tests
    @smoketest
    # This is an example comment
    Scenario: Product search works as expected
        Given I am on the Web shop homepage
        And I am logged in
        When I search for 'Packt'
        Then I get a list of items containing 'Packt'
```

Even though this feature file is taken from a website test (commonly called *UI test*), Gherkin can be used to describe any behavior of any system under test – potentially even cases that have nothing to do with software at all.

Let's take a look at the different parts of the preceding file.

Features

Gherkin files have the `.feature` extension and always start with a `Feature:` line followed by a description. Each feature file collects test scenarios that belong to a defined generic business case indicated by the given description. In our example, the feature name Web shop tests shows that this is a collection of test cases for our imaginary online store.

Scenarios

A feature can include one or more scenarios that are associated with it. Each scenario is a test case related to a specific business case, user journey, or behavior.

Scenarios are made up of different keywords and parts that we will look at in the following sections.

Tags

Within feature files that contain multiple scenarios, you can use one or more custom tags to further group them into test suites (in our case, the tag `@smoketest` could mean that the following scenario belongs to a set of small, basic tests that are run after a deployment to quickly verify that our application is usable). In the course of this book, you will see how these can be used to pre-select the tests you want to run at a specific point in your software development lifecycle.

Steps

Scenarios are made up of different steps that are executed from top to bottom. Each one describes either the state of the system under test, certain actions, or assertions. Their generic nature makes them excellent for reuse across multiple test cases.

> **Keywords can be substituted**
>
> From the processing of the scenarios by a Gherkin-based framework, it does not matter at all which keywords are used in which order. They are used solely for comprehensibility and semantic correctness. In any case, you should get into the habit of following these conventions.

Next, let's look at the different types of Gherkin keywords you can use.

The Given keyword

This keyword indicates the initial state and prerequisites of an application, its user, or data. In our example, `Given I am on the Web shop homepage` tells us where the users of the application start their journey in this test case. It is deliberately kept simple to increase comprehensibility and does not contain any technical details.

The When keyword

Any interactions with or changes to the system under test start with a When keyword. When I search for 'Packt' describes clearly *what* the users do and what their intention is but not *how* the webshop deals with this on a technical level. Later, we will see that Karate bends this core BDD rule and why it does it this way.

The Then keyword

The final part of most test scenarios is one or more assertions to verify a certain outcome or state of the system under test. This is expressed by Then as in Then I get a list of items containing 'Packt'. Again, this clearly states what outcome is expected from the system but not *how* exactly this is solved by the application.

The And keyword

The And keyword can connect multiple Given, When, or Then lines depending on if there is more than one prerequisite, action, or assertion. It is just "syntactic sugar" to make a scenario more readable.

In our example it connects two Given statements:

- The user is on the homepage and...
- ...the user is logged in

The But keyword

Like And, the But keyword can be used to connect steps. It is not as common as its counterpart but can express certain conditions better. Take the following, for example:

```
Given I am on the Web shop homepage

And I am not logged in
```

That can also be formulated as follows:

```
Given I am on the Web shop homepage

But I am not logged in
```

This would emphasize even more that the focus of this test is on an unauthorized user.

Catch-all steps (*)

Gherkin has one more step keyword that all different step types (Given, When, Then, And, and But) can be substituted with. A * step is mostly used when it does not really fit the natural flow of what should be tested but instead contains utility functions or setup code, or helps with debugging test scenarios.

An example is Karate's `print` function, which logs a string on the command line for simple tracing of values:

```
* print "The value of my variable is", myVariable
```

Here, it does not directly play a role in any part of the tested behavior. Instead, it can be considered as a utility step that still has to follow the chronological sequence of events but could easily be separated from them without affecting the overall test functionality. In Karate, an advantage of this is that these *bullet points* can be suppressed in the generated test reports, as we will later discover.

Comments

Comments can be used to make certain steps clearer, especially when they contain information that might be hard to understand for newer team members or colleagues that are not as familiar with the tested business domain.

Also, they can be used in case a test is temporarily deactivated to explain further why this was done or what needs to happen before it can be active again.

Comments can be placed anywhere in a Gherkin file. They start with a # symbol (like the `# This is an example comment` line in the example) and are ignored by Cucumber and Karate alike.

Additional Gherkin syntax

In this section, we only looked at the basic Gherkin keywords that will be necessary in the first steps toward writing test cases.

The Gherkin specification has some more advanced language constructs such as the following:

- **Data tables**: For specifying multiple sets of data for a step
- **Example tables**: For running scenarios with different sets of data
- **Background scenarios**: That define common steps in multiple scenarios
- **Scenario outlines**: For combining similar scenarios into one
- **Docstrings**: For more complex data definition in a step

All of these will play a role in later chapters when we look at code reuse, keeping tests free from code duplication, and the overall structuring and fragmentation of test suites

Writing good BDD scenarios

When doing BDD, test scenarios should typically have the following characteristics to keep them clear, understandable, and concise across your test suites:

- **Scenarios should be able to run independently**: This is important because depending on whether all or just a subset of scenarios is run or if they are run in multiple threads at the same time, we cannot always be sure of the order of execution. That means that if scenarios depend on each other (for example, one that creates a new user and one that deletes the new user), we can run into problems if both are executed at the same time or the wrong way around

- It makes sense to run scenarios in parallel early on to be sure that they can run without disturbing others.

- **Scenarios should be concise and easy to follow**: Your scenarios should not execute too many steps even though the Gherkin format is predestined for easy comprehension of the test cases. This is beneficial for a greater understandability and easier analysis of exactly what happened in case of an error.

- **Scenarios should not test too many different things at the same time**: This one is closely related to the second point – understandability. Commonly, a scenario fails as soon as a step fails, skipping all subsequent steps. Therefore, if a scenario tests too many things in succession, hints of faulty functions are lost.

- **Scenarios should not duplicate other scenarios**: In many cases, it might not make sense to do the same test repeatedly. For example, if you start multiple tests with the same request to the same API, you don't necessarily need to check that this API delivers the expected data format in every individual scenario. This might not always apply, for example, when similar tests belong to different test suites that might be run at different points in time.

- **Scenarios should not contain concrete technical information**: When writing behavior-driven tests, the focus is on high-level behaviors and how a system is interacted with. There should not be any references to implementation details or certain elements a user interacts with. The basic idea is to describe *what* should be done, but not *how*.

> **Karate is not true BDD**
>
> Almost all these points are valid tips for writing good Karate tests. However, the last guideline ("Scenarios should not contain concrete technical information") is one that we will revisit later in this chapter. Karate's idea is radically different in this regard, and we will see both why this makes sense and why the framework has moved away from BDD altogether.

Now that we have an idea about BDD and the Gherkin language, let's see in the next section how Gherkin statements can be linked to source code.

Glue code

So-called *glue code* builds the bridge between the Gherkin steps and the implementation of concrete test code (it *glues* these two together). This is how the test framework can execute the functions matching the respective step including the passed parameters.

Let's look at an example implementation of this part of the scenario:

```
Given I am on the Web shop homepage
When I search for 'Packt'
```

First, we will look at how Cucumber deals with glue code so that we can better compare it to Karate's approach afterward.

Cucumber glue code

The general mechanism of Cucumber-based frameworks is working by linking steps whose names fit predefined **Cucumber expressions** (https://github.com/cucumber/cucumber-expressions#readme) that are assigned to certain Java functions by annotations. In older versions of Cucumber, this matching of steps to glue code was done via **regular expressions** (https://en.wikipedia.org/wiki/Regular_expression), but Cucumber expressions turned out to be way easier to use when implementing glue code yourself:

```
Package blog.softwaretester.gherkin;
import io.cucumber.java.en.Given;
import io.cucumber.java.en.Then;
import io.cucumber.java.en.When;

public class StepDefinitions {
    @Given("I am on the Web shop homepage")
    public void goToHomepage() {
        System.out.println("Go to homepage");
    }

    @When("I search for {string}")
    public void search(final String searchTerm) {
        System.out.println("Search for " + searchTerm);
    }
}
```

The `goToHomepage()` function is invoked when Cucumber encounters a step definition that matches the `@Given("I am on the Web shop homepage")` annotation. It is good to know that, even though the `@Given` annotation is used here, this matches any other keyword, too, if the following text part is the same.

It is analogous to the step after that: `search(final String searchTerm)` is connected by the `@When("I search for {string}")` annotation. It is noteworthy that here, a dynamic parameter exists that can be any string (defined by the `{string}` placeholder) so that the following is true:

- This string parameter can be passed from the scenario step to the connected function (in our example, `searchTerm`)

- It can match any string so that here you avoid having to write a new function for each value – this ensures efficient code reusability

> **Demo steps**
> Please note that in the preceding method implementations, the steps do not have any concrete functionality except for printing out a message in order to show they are really being invoked.

When running the preceding example in Cucumber, we can see in the console log that this connection works as expected:

```
[INFO] Running blog.softwaretester.gherkin.RunCucumberTest
Go to homepage
Search for Packt
```

The complete example can be found at `https://github.com/PacktPublishing/Writing-API-Tests-with-Karate/tree/main/chapter01/cucumber-glue-example`.

Now that we know how Cucumber handles glue code, let's check out why Karate separated from Cucumber and implemented its own approach.

The Karate way

Karate uses essentially the same mechanism and concepts as Cucumber but with some interesting differences.

Up until version 0.9.0 of Karate, it was based on Cucumber because of its rich feature set and very good support in common **integrated development environments** (**IDEs**). However, Peter Thomas decided in 2018 to separate from Cucumber while keeping Karate fully compatible.

The main reasons for this decision were that Karate should not be dependent on the timeline, priorities, and feature set of Cucumber but instead have the freedom to move ahead with development at a faster pace. Basically, development is more gradual with the Cucumber framework because it must support and coordinate many different language bindings. Since Karate supports only one programming language, there is of course a speed advantage.

Also, this move allowed Karate to control all phases of the test lifecycle, including reporting, and not be tied into the existing Cucumber infrastructure, specifications, and technologies.

Karate, as we have seen, borrows Gherkin but explicitly does not use BDD. The reasoning behind this is that Karate tests are not specifically aimed at non-technical roles but instead at testers and developers with a more technical background.

Peter Thomas explains it like this:

> *Karate makes sense especially for "platform" teams creating and maintaining web-services, and where product-owner involvement in acceptance-testing is not the highest priority.*

Thomas, P. (2017). *Karate is not true BDD,* `https://medium.com/hackernoon/yes-karate-is-not-true-bdd-698bf4a9be39`.

Using Gherkin as a test DSL is intended to simplify testing and not just bluntly repeat what other frameworks already provide anyway.

Now, let's look at an example of how Karate handles glue code internally.

Karate glue code

Karate provides common predefined steps plus the matching glue code directly inside the framework. That means that you do not have to write any code in many cases since the steps you need are already implemented.

Some steps go way beyond the simple conversion of string arguments into primitive method parameters but instead process those further. This makes it possible to use steps that contain JavaScript, **JSONPath** or **XPath** expressions, Java function calls, and more. We will see the details of this approach when we deal with this in *Chapter 3, Writing Karate Tests*.

We can check how this is done by looking at the internal implementation of a scenario step, in this case, Karate's `print` statement:

```
* print "The value of my variable is", myVariable
```

In Karate's `ScenarioActions` class (available in its GitHub repository `https://github.com/karatelabs/karate`), we find this piece of code that connects the preceding scenario step to the glue code:

```
@Override
@When("^print (.+)")
public void print(String exp) {
    engine.print(exp);
}
```

We can clearly see that this is done in analogy to Cucumber. The only differences are that this implementation is predefined and uses a regular expression instead of a Cucumber expression. In this example, the statement `"^print (.+)"` means find a line starting with *print* (`^print`), take everything that follows, and interpret it as a parameter (`(.+)`). Like in a Cucumber expression, the parameter is then passed on to a function that can process it.

> **Note**
>
> This was just an example to illustrate the inner workings of the framework. You will most likely never be in touch with this kind of code again when authoring software tests with Karate!

The downside of this approach of embedding glue code within the framework should not remain unmentioned. It basically means that this functionality is locked away and cannot be changed anymore by the framework users. Also, adding your own glue code is not possible like it would be in a Cucumber-based framework. Not only is this not available, but it is also not desired by the authors of the framework, as this could violate Karate's conventions and lead to unforeseen errors. Since Karate's step implementations are very flexible as they are, they can potentially cover just about any use case without ever needing to be extended with custom glue code anyway.

The overall goal is to keep the framework simple and consistent and limiting the possibilities can be seen as an advantage here.

In this chapter, we covered BDD, the Gherkin language elements, and its technical implementation in Cucumber and Karate. Also, we talked about why and how Karate deviated from Cucumber in some core ideas. Now, let's look at the different data types that Karate can work with and how this makes it a big strong point for this framework.

Supported data types in Karate

One of the strengths of the Karate framework is its native support for a variety of different data types that are typically associated with API payloads and responses. That means that it is possible to use them in their original form within test scripts without the need of escaping certain characters or converting them to be usable.

Let's look at some widely used data formats here and the way they would represent data about this book and the first chapter for comparison.

JSON

JSON is pretty much the standard exchange format for REST APIs. Additionally, it is the standard format for **GraphQL**. It is small, structured, and human-readable while being a great format to work with JavaScript because it can be used natively there. This is a simple JSON example that shows how data can be represented by nesting keys and values:

```
{
    "book": {
        "title": "Writing tests with Karate",
        "author": "Benjamin Bischoff",
        "chapters": [{
            "number": "1",
            "title": "Karate's core concepts"
        }]
    }
}
```

This is the format we will encounter most throughout the book, as Karate can also represent other formats through JSON and make them testable.

GraphQL

GraphQL was developed by Facebook and released to the public in 2015. In 2018, it was transferred to the newly formed GraphQL foundation under the Linux Foundation (https://www.linuxfoundation.org/).

GraphQL's request and response format is JSON. Contrary to a REST API that always returns fixed data structures, users can define exactly what data they need in each request and only the requested data is then returned. This saves bandwidth and enables fetching deeply nested data without making multiple requests in a row. From a structural point of view, the only thing different from JSON responses is the added data object that wraps the expected JSON response:

```
{
    data: {
        "book": {
            "title": "Writing tests with Karate",
            "author": "Benjamin Bischoff",
```

```
        "chapters": [{
            "number": "1",
            "title": "Karate's core concepts"
        }]
    }
  }
}
```

In *Chapter 6, More Advanced Karate Features*, we will check out how to handle this within the Karate framework.

XML

Like JSON, **Extensible Markup Language** (**XML**) is another human-readable format but with some more overhead for the same set of data. Karate handles this format equally well, though:

```xml
<?xml version="1.0" encoding="UTF-8" ?>
<book>
    <title>Writing tests with Karate</title>
    <author>Benjamin Bischoff</author>
    <chapters>
        <chapter>
            <number>1</number>
            <title>Karate&#x27;s core concepts</title>
        </chapter>
    </chapters>
</book>
```

You can see that XML is more redundant than JSON because it uses similarly named opened and closed tags (e.g., `<title>` and `</title>`) to enclose values.

YAML

Yet another markup language (**YAML**) has a very minimalistic, indentation-based syntax. It is a rather new format that gained popularity mainly for its use in configuration files. The book example would look as follows, making this the most readable (but also error prone) format of them all:

```yaml
---
book:
    title: Writing tests with Karate
```

```
author: Benjamin Bischoff
chapters:
    - number: 1
      title: Karate's core concepts
```

Karate can work with the YAML format easily by automatically converting it into JSON!

CSV

Comma-separated values (CSV) files contain rows of data that are separated by line breaks. Each line is a dataset (or record) in which values are typically delimited by commas. It is often used as a data exchange format from and to spreadsheets, such as Excel:

```
title,author,chapter/number,chapter/title
Writing tests with Karate,Benjamin Bischoff,1,Karate's core
concepts
```

Like YAML, Karate converts CSV into JSON arrays automatically!

Other text-based formats

Luckily, Karate supports all textual formats. For those that are not included natively, we are free to use generic strings. However, depending on how obscure the format in question is, we need to write our own Java or JavaScript handling for such edge cases.

Binary formats

Through Karate's Java interoperability, it is possible to handle virtually any binary file format. Additionally, Karate has the special `bytes` type to convert any binary data into standard byte arrays to use as payloads.

If you need to use `websocket` connections that use binary data, there is even a dedicated built-in method available, `karate.webSocketBinary()`.

In this section, we have seen the different data formats that Karate can understand and work with. You can probably already see how powerful this framework is. Next, we will check how Karate's JavaScript engine helps even further with this.

The JavaScript engine

Karate's JavaScript engine is what really distinguishes this framework from many others. It is built on GraalVM (`https://www.graalvm.org/`), which allows so-called **polyglot programming**. This means that it can mix and match different programming languages together in a single application and even pass values back and forth. This makes it possible to use native JavaScript code both directly embedded into feature files, within Karate's main configuration, or even stored in external files.

The direct integration of JavaScript in Karate is especially helpful when working with JSON and similar formats. This format is a first-class citizen in JavaScript whereas in Java you would need additional libraries to effectively work with it (for example, `Gson` or `Jackson`). This enables test authors to directly write JSON in steps without the need to escape or encode certain characters. It also allows for more straightforward assertions and matching by the direct integration of JSONPath as we will see further on.

In *Chapter 7, Extending Karate Functionality*, we will explore how JavaScript can be used to customize and simplify more complex validations, matching, or data manipulations.

The JavaScript engine is only one of the parts that power Karate's internals. Let's now look at the other one: Java itself.

Java interoperability

As we have already learned, despite Karate's close relationship with JavaScript, it is purely written in Java. This ensures that it is fast, concise, and runs on virtually any local system, CI/CD servers, and clouds. It works perfectly with the Maven build system, which we will use in the next chapter to set up our Karate project. Java also offers mature development tools, extensions, and libraries to enable rapid implementation.

Running unit tests is also very easy and tightly integrated with virtually all common IDEs so that we can check test results directly there without switching between tools.

Additionally, the direct Java access allows for some interesting functionalities that open a whole new world in terms of Karate testing. Since Java has tons of specialized libraries and built-in features for common business-related tasks (for example, database access through Hibernate), we can use these directly. This is a great example of the "use the right tool for the job" mantra – we can use JavaScript or Java, whichever can handle the task better!

As we will explore further in *Chapter 7, Extending Karate Functionality*, Karate can call Java methods and utilities via JavaScript, which makes this a very powerful and easy-to-extend feature.

Summary

In this chapter, we talked about why the Karate framework was developed and the many features and possibilities it has that help write expressive yet simple test cases. Additionally, we learned about the keywords and structure of Gherkin files and explored why Karate adapted this language but not the BDD standard itself.

In *Chapter 2, Setting Up Your Karate Project*, we will look at how to set up the development environment to be fit for Karate and the various ways to start a new Karate project from scratch using Karate standalone and Maven.

You will learn step-by-step how to configure the necessary environment and IDE plugins to streamline your development experience and make testing fun. Also, we will look at some unique features that are only available in the Karate plugins for **Visual Studio Code** and **IntelliJ IDEA**.

2

Setting up Your Karate Project

Karate allows us to use it in a variety of ways. We could use the standalone version of it to just have a simple way of running tests and trying out its capabilities. However, it is more convenient to set it up with a proper build tool and **Integrated Development Environment** (**IDE**), as this greatly simplifies development and facilitates the running and debugging of tests.

We will check out how to get your system prepared for the Karate framework and look at different build tools and IDE plugins. By the end of this chapter, you will understand what is needed to develop and run Karate tests locally on your system and be prepared for the next chapters, where we will dive deeper into the Karate DSL.

In this chapter, we will cover these main topics:

- Installing Java
- Getting to know Karate standalone
- Preparing the IDE
- Setting up Karate projects with **Maven**
- The roles of the different Karate project files

Let's look at the technical requirements for the following sections next.

Technical requirements

The code examples for this chapter can be found at `https://github.com/PacktPublishing/Writing-API-Tests-with-Karate/tree/main/chapter02`.

Disclaimer

In theory, you could use any text editor to write Karate tests, since they are simple text files at their core. However, using a real IDE makes development easier and more comfortable. We will not use any paid tools within this book in order to make this content available to you without any limitations.

You will require the following in this chapter:

- A **Java Development Kit** (**JDK**) to develop and run tests (Karate requires version 8 or higher). We will see in the next section how to install it.

- **Maven** to manage the dependencies of our Karate projects (it is possible to use Gradle as well, but we will mostly use Maven throughout this book, since setting up Karate projects is more streamlined). We will discuss how to install and set up Karate projects using Maven further on in this chapter.

- **Visual Studio Code** or **IntelliJ IDEA** as the IDE (we will use Visual Studio Code throughout this book, since it has a dedicated, free Karate plugin with a lot of helpful features).

- **Chrome browser** to try out the web UI automation examples later (`https://www.google.de/chrome/`).

> **Operating system differences**
>
> Karate runs on all major operating systems. All successive steps are rather similar. The most dramatic difference is how programs are installed and that Linux and macOS use Terminal, while Windows uses Command Prompt or the PowerShell window.
>
> For demonstration purposes, I will use Windows and its default **standalone Command Prompt**. When in Visual Studio Code, I will use the built-in **PowerShell** window (called **Terminal**) instead. On macOS, Visual Studio Code's terminal window uses the system terminal, which internally uses a **Unix shell**.

In the next section, we will look at the first necessary step to prepare our system – how to install Java.

Installing Java

For the development of Karate tests, it is beneficial to have a JDK installed on your system. This allows you to not only run but also create and compile Java programs. Regardless of what IDE you use, this is a crucial component for effective Karate development.

You can download Java for your system here: `https://www.oracle.com/java/technologies/downloads`.

> **SDKMAN!**
>
> For Unix-based systems, **SDKMAN!** is a great alternative to easily install JDKs as well as Maven or Gradle. It sets the necessary system environment variables and even allows installing multiple ones and switching between them. You can check it out here: `https://sdkman.io`.

In principle, it doesn't matter which version you choose as long as it is Java 11 or above. I selected Java 17 here, because it is the so-called **Long-Term Support (LTS)** version and is supported until at least September 2024.

> **Which Java package?**
>
> It makes sense to choose the Java installer package for your operating system instead of the ZIP release, since it is easier to register on your operating system.

After installation, you can verify that Java is correctly installed by opening a Terminal window or Windows Command Prompt and typing `java  -version`, followed by *Enter* key. This should answer with the version of Java you just installed:

```
PS C:\Users\bbischoff> java -version
java version "17.0.4.1" 2022-08-18 LTS
Java(TM) SE Runtime Environment (build 17.0.4.1+1-LTS-2
Java HotSpot(TM) 64-Bit Server VM (build 17.0.4.1+1-LTS-2,
mixed mode, sharing)
```

On my screen, it looks like this, indicating that Java is available and correctly launchable.

Setting the JAVA_HOME environment variable

JAVA_HOME is a system-wide environment variable that contains the path to your Java installation. For later purposes, it may be important to set this so that Java can be correctly found from any path and IDE, especially when using Maven.

Depending on your setup, the following step might not be necessary, but I will describe it here nonetheless if there are any issues.

```
Please set the JAVA_HOME variable in your environment to match the
location of your Java installation.
```

Figure 2.1 – An error message if there is a missing JAVA_HOME

Next, let us see how to set up this environment variable in Windows.

Setting JAVA_HOME in Windows

The following steps are required to set JAVA_HOME in Windows:

1. If you enter path in the Windows search, you should find the **Edit the system environment variables** entry.

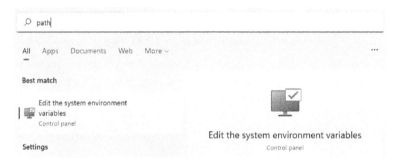

Figure 2.2 – Finding the system environment variables settings

2. Click on it to open the **System Properties** dialog. Switch to the **Advanced** tab and click on the **Environment Variables…** button.

3. Here, you can enter JAVA_HOME as the variable name. The variable value should be your root JDK directory.

Figure 2.3 – Setting JAVA_HOME

4. You can use the **Browse Directory…** button to choose the correct folder – on Windows, it is typically under Program Files | Java.

Figure 2.4 – Browsing to the JDK folder

After clicking **OK**, the JAVA_HOME environment variable should be available in your system. You can quickly test this by opening Command Prompt, typing in echo %JAVA_HOME%, and hitting *Enter*.

When you see the path that you entered in *step 3*, your JAVA_HOME variable is set up correctly:

```
C:\Users\bbischoff>echo %JAVA_HOME%
C:\Program Files\Java\jdk-17.0.4.1
```

Now that we have set up JAVA_HOME in Windows, let's quickly check the same for macOS and Linux.

Setting JAVA_HOME in macOS

For macOS, it might not be necessary at all to set JAVA_HOME. In case you do get a warning, though, I will quickly describe how to do it for **macOS Catalina** (10.5) and above (for macOS systems older than 10.5, you should replace .zshenv in the following commands with .bashenv):

1. Open the macOS terminal.app.
2. Open the ~/.zshenv file in the **nano** command-line text editor by typing nano ~/.zshenv, followed by *Enter*. This is the standard configuration file for environment variables for the default macOS **zsh** Terminal shell.
3. Add the following line at the end of the .zshenv file. This sets the JAVA_HOME variable to /usr/libexec/java_home, which points to your default JDK path:

    ```
    export JAVA_HOME=$(/usr/libexec/java_home)
    ```

4. Press *CTRL + X* to exit nano and *Y* to accept the changes.

5. Reload the configuration by typing `source ~/.zshenv` and pressing *Enter*.

Now, when you type `echo $JAVA_HOME`, you should see the correct JDK path. On my screen, it looks like this:

```
echo $JAVA_HOME
/Users/bbischoff/Library/Java/JavaVirtualMachines/
openjdk-17.0.1/Contents/Home
```

Next, let's learn how to set this up on a Linux machine.

Setting JAVA_HOME in Linux

For Linux, the steps are very similar to the macOS approach. The main differences are that Linux does not have the `/usr/libexec/java_home` path that automatically contains the path to the current JDK, and that the Terminal shell configuration is in `~/bash_profile`.

So, you should add the following line to the end of the configuration file like this (you need to use the correct path to the JDK):

```
export JAVA_HOME=path to your JDK
```

When you open a new Terminal window and type `echo $JAVA_HOME`, this should answer with the path you specified, like the macOS example we saw earlier.

In this section, we looked at how to install the JDK and how the `JAVA_HOME` variable can be set on different operating systems.

Now that we have a working Java set up, let's look at Karate's simplest way to get started, Karate standalone.

Getting to know Karate standalone

`Karate standalone` is a single Java JAR file that contains all parts of the framework. It can be executed on the command line and just needs Java installed. This makes it possible to check out its capabilities without having to install an IDE.

Another interesting use case is its ability to spin up a mock server. We will see more about this in *Chapter 6, More Advanced Karate Features*.

> **Karate standalone as a test runner**
>
> The standalone version can be used within Visual Studio Code to run tests directly from the IDE. We will not use it for this purpose, though, as we will deal with a more real-world setup throughout the following chapters.

Downloading Karate standalone

The Karate Standalone download is available on the Karate releases GitHub site: https://github.com/karatelabs/karate/releases

Make sure you download the **ZIP Release option**, which can be found under the latest Karate version (the release candidate versions marked with **RC** in the version number might not include a standalone release).

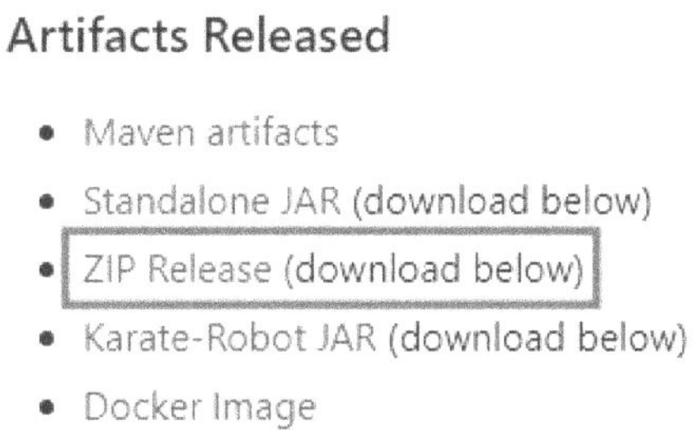

Artifacts Released

- Maven artifacts
- Standalone JAR (download below)
- ZIP Release (download below)
- Karate-Robot JAR (download below)
- Docker Image

Figure 2.5 – Downloading Karate standalone

Once downloaded, unzip it to a convenient directory, open a Terminal window, and navigate to the Karate folder. In the root directory, there are two start scripts – karate.bat for Windows and karate for Linux and macOS. If you type karate (Windows) or ./karate (Linux/macOS) and press *Enter*, you should see output similar to that shown in *Figure 2.6*.

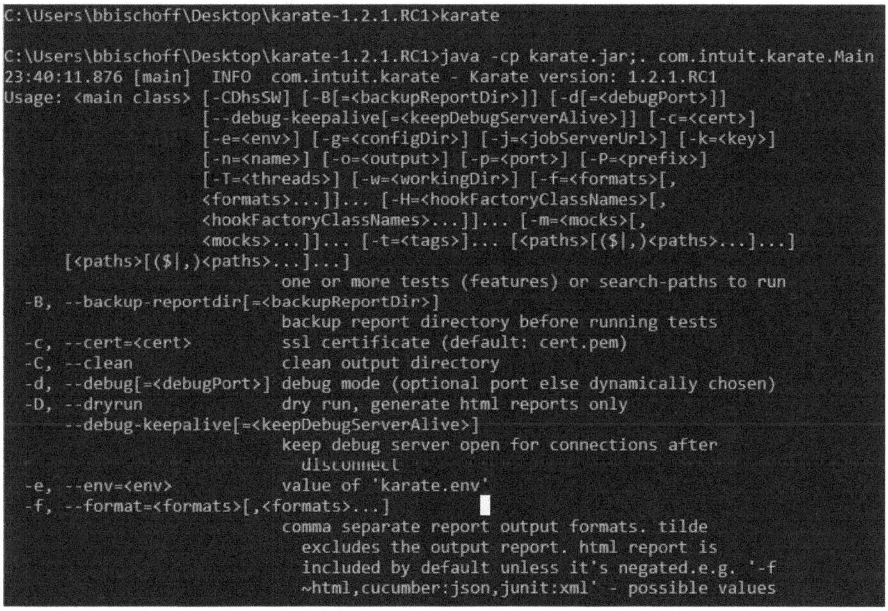

```
C:\Users\bbischoff\Desktop\karate-1.2.1.RC1>karate

C:\Users\bbischoff\Desktop\karate-1.2.1.RC1>java -cp karate.jar;. com.intuit.karate.Main
23:40:11.876 [main]  INFO  com.intuit.karate - Karate version: 1.2.1.RC1
Usage: <main class> [-CDhsSW] [-B[=<backupReportDir>]] [-d[=<debugPort>]]
                    [--debug-keepalive[=<keepDebugServerAlive>]] [-c=<cert>]
                    [-e=<env>] [-g=<configDir>] [-j=<jobServerUrl>] [-k=<key>]
                    [-n=<name>] [-o=<output>] [-p=<port>] [-P=<prefix>]
                    [-T=<threads>] [-w=<workingDir>] [-f=<formats>[,
                    <formats>...]]... [-H=<hookFactoryClassNames>[,
                    <hookFactoryClassNames>...]]... [-m=<mocks>[,
                    <mocks>...]]... [-t=<tags>]... [<paths>[($|,)<paths>...]...]
    [<paths>[($|,)<paths>...]...]
                        one or more tests (features) or search-paths to run
  -B, --backup-reportdir[=<backupReportDir>]
                        backup report directory before running tests
  -c, --cert=<cert>       ssl certificate (default: cert.pem)
  -C, --clean             clean output directory
  -d, --debug[=<debugPort>] debug mode (optional port else dynamically chosen)
  -D, --dryrun            dry run, generate html reports only
      --debug-keepalive[=<keepDebugServerAlive>]
                        keep debug server open for connections after
                        disconnect
  -e, --env=<env>         value of 'karate.env'
  -f, --format=<formats>[,<formats>...]
                        comma separate report output formats. tilde
                        excludes the output report. html report is
                        included by default unless it's negated.e.g. '-f
                        ~html,cucumber:json,junit:xml' - possible values
```

Figure 2.6 – Karate standalone output

This is a detailed list of all the options and flags you can specify when launching Karate standalone. As I just want to demonstrate its basic usage, we won't be going through everything mentioned here but instead take a quick look at how to run an example test scenario with it.

JBang versus JRE versus JDK

In the Karate manual, **JBang** (`https://www.jbang.dev`) is recommended to spin up a Java test environment for playing around with Karate standalone. This is a tool to run Java applications such as Karate with a minimal setup.

Since we already have the JDK installed on our system, we don't need to use JBang.

If you want to run tests on a different system later where you don't develop Karate tests, it only needs a **Java Runtime Environment** (**JRE**) and not the full-blown JDK.

Running an example test with Karate standalone

`Karate standalone` comes with everything you need to try out the different flavors of tests that are possible with it. In the `src/demo` folder, you can find the following examples:

- `src/demo/api` contains an example for API tests
- `src/demo/mock` includes an example of Karate acting as a mock server
- `src/demo/robot` contains a Robot example for testing desktop applications (there is a dedicated setup page for this if you want to try it out later: `https://github.com/karatelabs/karate/wiki/Karate-Robot-Windows-Install-Guide`)
- `src/demo/web` includes a web UI test example (you need to have **Google Chrome** installed for it to run properly!)

To run the examples, you can specify the scenario on the command line.

For example, if you want to run the web UI test, you can execute this command on Windows:

```
karate src/demo/web/google.feature
```

On macOS and Linux, it looks like this:

```
./karate src/demo/web/google.feature
```

If everything is set up correctly, this will result in the test running right on the command line, and you will see a Chrome browser window open.

Firewall warning

Since this test is communicating with your network, a firewall warning may come up on the first run.

Figure 2.7 – Firewall warning

We will leave `Karate standalone` on the system for now to reuse it and its example projects at a later stage. Feel free to run the other demo scenario as well if you want.

Figure 2.8 – Running a Karate standalone UI test

In this section, we checked out `Karate standalone` and how we can run some sample tests with it. Next, let's set up an IDE to create a more user-friendly environment to develop tests during the following chapters.

Preparing the IDE

It is best to use a proper Java IDE to develop Karate tests. In this section, we will look at how to set up **Visual Studio Code** and **IntelliJ IDEA Community Edition** – both are officially supported by Karate Labs.

> **The IDE used in the book**
>
> Throughout the course of the book, the examples will use Visual Studio Code. However, the examples are, of course, runnable in any other IDE that supports Java.

Setting up Visual Studio Code

Now, we will set up the Visual Studio Code IDE, also known as VS Code. It is a free IDE from Microsoft that can be used for a variety of programming languages (including Java and JavaScript) and supports a lot of different plugins.

Downloading VS Code

You can download VS Code from here: `https://code.visualstudio.com`. It is advisable to download the **Stable Build** and not the **Insiders Edition**, as this is a preview release and may not be stable.

After the installation, you will be guided with a **Get Started with VS Code** assistant, where you can choose a color theme and get to know some of VS Code's features.

Importing the example project

We will use the Karate standalone example project to set up Karate support and verify our setup, as we did with the IntelliJ IDEA setup:

1. In VS Code's welcome screen, click the **Open Folder…** button to bring up the open dialog.

Figure 2.9 – The VS Code welcome screen

2.   Choose the unzipped Karate standalone folder and confirm it with **Select Folder**.

Figure 2.10 – Selecting the Karate standalone project

> **Trusting the project**
>
> VS Code might ask you whether you trust the project. In this case, you should choose **Yes, I trust the authors** to be able to work with it. Otherwise, it is only possible to view its code.

After VS Code imports the folder, you will be able to see its contents in the explorer panel on the left side. The demo feature files are located inside the `src` directory.

Figure 2.11 – VS Code with the opened project

Next, let's look at how to install the Karate plugin so that we can work more efficiently with feature files.

Installing the Karate plugin

When opening the project and selecting one of the feature files, there will be no syntax highlighting yet. Luckily, VS Code will prompt you if you want to search its Marketplace for a plugin to match files with a `.feature` extension.

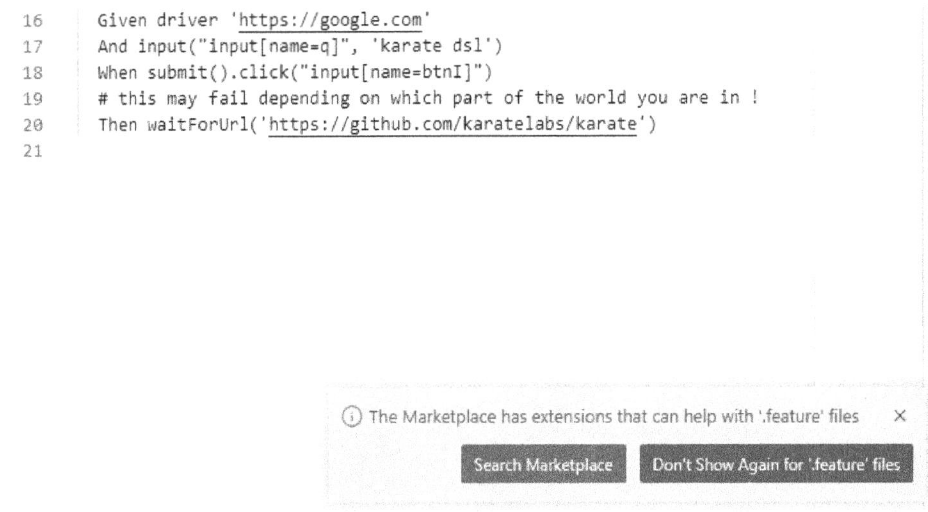

```
16    Given driver 'https://google.com'
17    And input("input[name=q]", 'karate dsl')
18    When submit().click("input[name=btnI]")
19    # this may fail depending on which part of the world you are in !
20    Then waitForUrl('https://github.com/karatelabs/karate')
21
```

ⓘ The Marketplace has extensions that can help with '.feature' files   ✕

Search Marketplace   Don't Show Again for '.feature' files

Figure 2.12 – The VS Code Marketplace prompt

If you click on the **Search Marketplace** button, VS Code will list matching plugins. We will use the open source **Karate Runner** plugin by Kirk Slota (the source code can be found at https://github.com/kirksl/karate-runner).

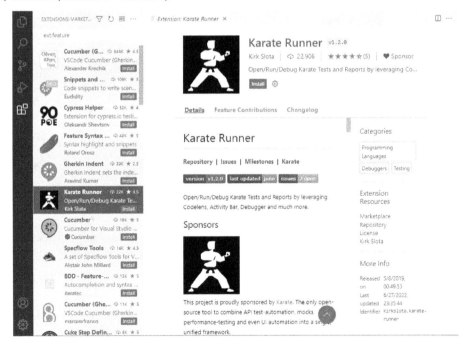

Figure 2.13 – VS Code Marketplace prompt

It is also possible to install extensions later via the **Preferences | Extensions** menu.

Open Recent	>	Extensions	Ctrl+Shift+X
Add Folder to Workspace...		Keyboard Shortcuts	Ctrl+K Ctrl+S
Save Workspace As...		Migrate Keyboard Shortcuts from...	
Duplicate Workspace			
Save	Ctrl+S	Configure User Snippets	
Save As...	Ctrl+Shift+S	Color Theme	Ctrl+K Ctrl+T
		File Icon Theme	
Auto Save		Product Icon Theme	
Preferences	>	Turn on Settings Sync...	

Figure 2.14 – VS Code installing extensions

After its installation, there will be an additional Karate tab, visible next to the explorer panel.

Figure 2.15 – The VS Code Karate tab

The Karate Runner plugin versus the Karate plugin

Even though there is an official Karate Labs VS Code plugin available (see `https://marketplace.visualstudio.com/items?itemName=karatelabs.karate`), we will use the free **Karate Runner**. As I write this, this free extension by Kirk Slota is more suitable for Karate development and supports features that you must pay for in the official extension. If you like, you can, of course, use the Karate Labs plugin instead. Be aware though that you will not be able to use the debugging features that are included for free with Karate Runner. These will be discussed further in *Chapter 4, Running Karate Tests*.

Also, the opened feature file should now have syntax highlighting, as seen in the following figure:

```
EXPLORER                    ···    ≡ google.feature  ×
∨ KARATE-1.2.1.RC1    ⊡ ⊡ ↻ ⊟       src > demo > web > ≡ google.feature > ☺ Feature: web-browser automation
  > .idea                                   Karate: Run | Karate: Debug
  > out                          ⊀  1    Feature: web-browser automation
  ∨ src\ demo                       2        for help, see: https://github.com/intuit/karate/wiki/ZIP-Release
    > api                           3
    > mock                         4    Background:
    > robot                        5        * configure driver = { type: 'chrome' }
    ∨ web                          6
      ≡ google.feature                     Karate: Run | Karate: Debug
  > target                     ⊀  7    Scenario: try to login to github
  ◆ .gitignore                      8        and then do a google search
  ≡ karate                          9
  ▓▓ karate.bat                   10    Given driver 'https://github.com/login'
  ▓ karate.jar                    11    And input('#login_field', 'dummy')
                                  12    And input('#password', 'world')
                                  13    When submit().click("input[name=commit]")
                                  14    Then match html('.flash-error') contains 'Incorrect username or passwo
                                  15
                                  16    Given driver 'https://google.com'
                                  17    And input("input[name=q]", 'karate dsl')
                                  18    When submit().click("input[name=btnI]")
                                  19    # this may fail depending on which part of the world you are in !
                                  20    Then waitForUrl('https://github.com/karatelabs/karate')
                                  21
```

Figure 2.16 – The feature's syntax highlighting

Other than in IntelliJ IDEA, VS Code should detect the installed JDK automatically, so we do not need to set this up explicitly here.

Setting up the VS Code plugin

In order to work with Java and Maven projects effectively, we need to add the Java extension for VS Code. Click on **File** | **Preferences** | **Extensions** or select the **Extensions** tab in the explorer. If you search for Java, the first plugin that comes up should be the **Extension Pack for Java**. This includes several plugins related to Java and Maven that are installed in one go. Just click the **Install** button and let VS Code do its work.

Figure 2.17 – The VS Code Extension Pack for Java

Settings

All the settings, installations, and modifications we have made so far are also accessible through VS Code's **File** | **Preferences** | **Settings** and **File** | **Preferences** | **Extensions** menus!

Please keep in mind that I will use VS Code in the following chapters because of its superior free Karate plugin. If you want to use IntelliJ IDEA instead of VS Code (or you already use it for other development tasks), the next section shows you how to do this.

Setting up IntelliJ IDEA

IntelliJ IDEA is specialized in features aimed at programming in Java. Through its extensive plugin infrastructure and customization options, it is well suited for Karate test development. However, it does not support JavaScript development in the free version, which is a small drawback when working with Karate's JavaScript extension.

Downloading IntelliJ IDEA

You can download IntelliJ IDEA from here: `https://www.jetbrains.com/idea/download`. Make sure you download the Community Edition, which is completely free.

Figure 2.18 – The IntelliJ IDEA Community Edition

When you launch IntelliJ IDEA after the installation, it greets us with a welcome dialog. Here, it is possible to make some adjustments, select a color scheme, and – what is most important for us – install plugins.

Installing the Karate plugin

Now, we can install the official IntelliJ IDEA Karate plugin from Karate Labs by going to **Plugins**, selecting **Marketplace**, and entering `karate` in the search field.

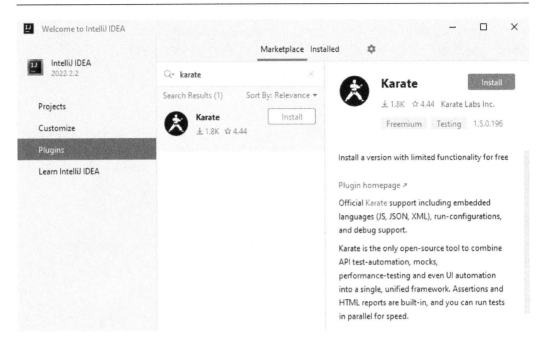

Figure 2.19 – The IntelliJ IDEA welcome dialog

Clicking either of the two **Install** buttons will show a warning, since the plugin is not made by IntelliJ IDEA. Click on **Accept** to install it anyway.

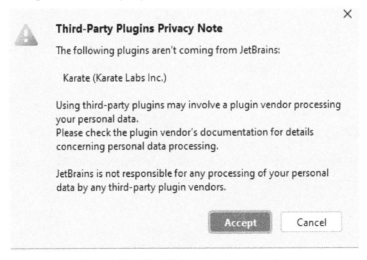

Figure 2.20 – The third-party plugin warning

After accepting it, the installation will finish, and the IDE needs to be restarted – this can be done by clicking on the **Restart IDE** button, as shown here.

Figure 2.21 – IDE restart after plugin installation

IntelliJ IDEA will now restart and automatically enable this plugin. Since we are using the **Freemium** (meaning unpaid basic) plugin version here, we will get essential features such as syntax coloring and code formatting but no auto-completion of our code.

It is possible to purchase a monthly subscription plan to get advanced features. If you want to know more about this, you can check out the information about features and pricing at https://www. karatelabs.io/plugins.

> **Settings**
>
> All settings, installations, and modifications we have made so far are also accessible through IntelliJ IDEA's **File | Setting** menu later!

Importing the example project

In order to test whether everything works, we can now import the Karate standalone example project into **IntelliJ IDEA**:

1. To do this, click the **Open** button in the welcome dialog.

Figure 2.22 – Opening a project through the welcome dialog

2. In the following dialog box, go to the Karate standalone folder and click **OK** to import it.

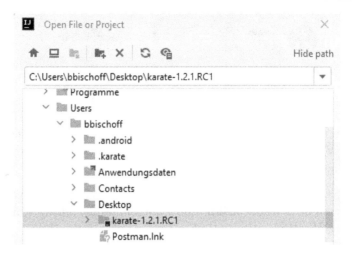

Figure 2.23 – Importing the Karate standalone example project

3. Alternatively, you can choose **File** | **New** | **Project from Existing Sources…** when you are inside the IDE.

Figure 2.24 – Opening a project through the file menu

> **Trusting the project**
>
> IntelliJ IDEA might ask you if you trust the project. In this case, you should choose **Trust Project** to be able to work with it. Otherwise, it is only possible to view its code.

After IntelliJ IDEA finishes importing the project, there is one more step we need to complete.

Assigning the JDK

In order to be able to run tests from the IDE later and get proper code completion, we need to assign the JDK to the project. Otherwise, you might get a warning later.

Figure 2.25 – The missing JDK warning

This is done from the **File | Project Structure** menu. Under the **Project** category from the left sidebar, you can then select the installed SDK (in my case, **17 Oracle OpenJDK**) and click **OK**.

Figure 2.26 – SDK selection

Now that the JDK is assigned to the project, we can continue and check whether everything is set up correctly.

Checking the setup

In the project view on the left side, all feature files should have a Karate icon, indicating that the plugin is active and feature files are correctly recognized. Double-click the `google.feature` file under `src/demo/web` to open it in the code window on the right.

Figure 2.27 – Opening google.feature

All Gherkin keywords (Feature, Scenario, Background, Given, When, Then, and And) in the opened feature files should be correctly highlighted now. Additionally, some Karate-specific commands will be highlighted as well. This means the Karate plugin is working as expected.

```
Feature: web-browser automation
    for help, see: https://github.com/intuit/karate/wiki/ZIP-Release

Background:
  * configure driver = { type: 'chrome' }

Scenario: try to login to github
  and then do a google search

Given driver 'https://github.com/login'
And input('#login_field', 'dummy')
And input('#password', 'world')
When submit().click("input[name=commit]")
Then match html('.flash-error') contains 'Incorrect username or password.'

Given driver 'https://google.com'
And input("input[name=q]", 'karate dsl')
When submit().click("input[name=btnI]")
# this may fail depending on which part of the world you are in !
Then waitForUrl('https://github.com/karatelabs/karate')
```

Figure 2.28 – The Feature file syntax highlighting

This concludes the installation of IntelliJ IDEA and the setup for working with the Karate framework. Next, let's check out how to create new Karate projects using Maven.

Setting up Karate projects with Maven

There are different ways to set up a Karate project; the most popular one uses **Apache Maven**. In the following sections, we will see how this works.

> IDE
>
> In the following sections, I will use VS Code. For IntelliJ IDEA, you can check out the basic steps here: https://www.jetbrains.com/idea/guide/tutorials/working-with-maven/creating-a-project.

Let's first quickly look at what Maven is used for.

What is Maven?

Maven is one of the standard tools in the Java world to simplify application creation and dependency management. It uses a central configuration file called pom.xml to describe the dependencies that a Java project needs and the steps to test, build, and deploy it.

> **This is not a Maven book**
>
> Maven is a very extensive project with a lot of different applications. In our case, we will use only a small subset of Maven, since we do not need to build and deploy applications but only manage dependencies and run tests.

Let's see how to install Maven in the next section.

Installing Maven

In order to use Maven to its full potential, it is necessary to install it on your system. Please look at https://maven.apache.org/download.cgi to find the correct file for your operating system. Unpack it into a directory of your choice.

Now, we must make this available on our system for VS Code to pick it up. In the following section, we will detail two different ways to do this.

Adding Maven to the PATH variable

Let's first look at how to set the system PATH variable so that it is accessible from anywhere on your system.

> **Note**
>
> This step might not be necessary, depending on how Maven is installed and how your specific version of the operating system handles this. If you prefer, you can try skipping this step and come back here if you are facing error messages about an unknown Maven installation later on.

Adding Maven to the system PATH variable in Windows

When Maven is added to the system's PATH variable, it is accessible from anywhere (any terminal window as well as VS Code). This can be done the same way we set other environment variables, as we did before with JAVA_HOME. Go to the **Environment Variables** window, select the **Path** variable, and click the **Edit…** button.

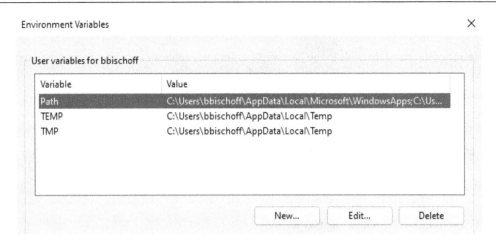

Figure 2.29 – Select the Path variable

This brings up the **Edit environment variable** dialog box, where you can add a new line by clicking **New**. Enter the path to your Maven `bin` folder by typing or using the **Browse…** button, and then click **OK** to approve.

Figure 2.30 – Adding the Maven bin folder

When typing `mvn  -v` now from any path in the terminal, it should be located correctly by the operating system. You can verify this by checking the printed version information about Maven, Java, and the operating system itself:

```
C:\Users\bbischoff>mvn -v
Apache Maven 3.8.6 (84538c9988a25aec085021c365c560670ad80f63)
Maven home: C:\Users\bbischoff\Desktop\apache-maven-3.8.6-bin\
apache-maven-3.8.6
Java version: 17.0.4.1, vendor: Oracle Corporation, runtime:
C:\Program Files\\jdk-17.0.4.1
Default locale: en_US, platform encoding: Cp1252
OS name: "windows 11", version: "10.0", arch: "amd64", family:
"windows"
```

Let us now look at how we can do the same in macOS.

Adding Maven to the system PATH variable in macOS

Like we did with JAVA_HOME earlier, we can use the nano ~/.zshenv command to edit the Terminal configuration file (or nano ~/.bash_profile on macOS operating systems older than 10.5). Now, add the following line to the end of the file. Replace your_maven_bin_path with the actual path to the Maven bin folder!

```
export PATH="your_maven_bin_path:${PATH}"
```

This command adds the Maven bin path to the existing PATH variable as a new entry and makes it available to your operating system by the export command.

If you open a new Terminal window afterward and type echo $PATH, you should see your Maven bin path as the first entry of your PATH variable.

Also, the mvn -v command should give you a similar output, as noted in the *Adding Maven to the system PATH variable in Windows* section earlier.

Adding Maven to the system PATH variable in Linux

For Linux, this works analogous to macOS, only that you need to add the line to ~/bash_profile instead of ~/.zshenv. The rest of the steps are the same as in the *Adding Maven to the system PATH variable in macOS* section.

Adding the Maven path to VS Code only

If you don't need Maven to be available system-wide but just in your IDE, you can also configure it only for VS Code. For this, open the VS Code settings under **File | Preferences | Settings** and type maven into the search bar.

Figure 2.31 – Setting the Maven path in VS Code

Select **Maven for Java** and look for the **Maven › Executable: Path** input field. Insert the full path to your Maven folder, including /mvn here, and close the settings dialog. Now, VS Code should be able to find your Maven installation.

You can verify this by opening the VS Code terminal panel under **View | Terminal** and typing mvn -v. It should print out the version information about Maven, Java, and the operating system. If this does not work right away, a restart of VS Code may be required.

Now that we know various ways to set up the Maven path, either system-wide or for VS Code only, let's see how to set up a new Karate project using the Maven archetype.

Setting up a Karate project with the Maven archetype

If the Extension Pack for Java is installed correctly, VS Code will show you a Java panel in the explorer when no project is currently open (you can use **File | Close Folder** if one is currently open). You might also need to click the **Explorer** icon again afterward to bring up the panel. This will give you the option to create a Java project, but what we need is the **Maven** tab. Clicking on the + icon will start the Maven wizard.

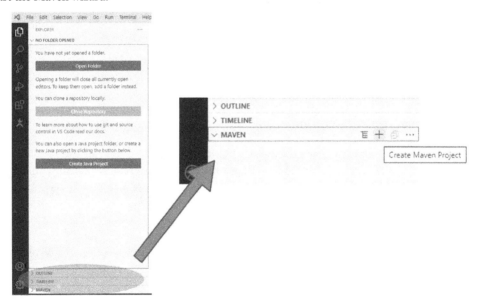

Figure 2.32 – Create Maven Project

You can also find the **Maven** tab at the bottom of the explorer when another project is open.

Next, we can select an **archetype**. This is essentially a project template that helps us to get started quickly. However, we only see some of the most used ones here. Clicking on **More...** will open the search to find more archetypes – including the Karate one.

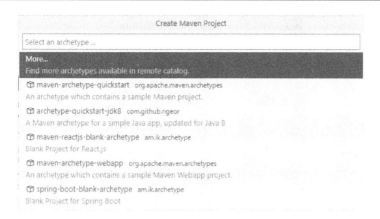

Figure 2.33 – Displaying more archetypes

Enter `karate` into the search box to find the **karate-archetype** option and select it.

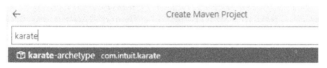

Figure 2.34 – Selecting the Karate archetype

The next step lists all available versions of available Karate archetypes; unfortunately, these are not sorted in a way that makes it easy to see the newest one. You can either go through the list or quickly check the newest version here: `https://mvnrepository.com/artifact/com.intuit.karate/karate-archetype`.

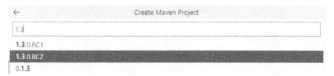

Figure 2.35 – Selecting the newest archetype version

Select the version you want to have (in my case, **1.3.0.RC2**). The next step consists of specifying a group ID for your project. This is a unique identifier for your project that usually starts with a reversed domain that you or your company owns (for example, my domain, `www.softwaretester.blog`, becomes `blog.softwaretester`). It does not have to comply with this standard if you don't have a domain of your own.

Figure 2.36 – Specifying the project group ID

Next, you can choose an artifact ID that usually represents the name of a resulting JAR file. However, since we are not using this project to compile a project but just to run tests with it, this can be anything. For demonstration purposes, I named mine karate-maven. Press *Enter* again to go to the last step of the Maven assistant.

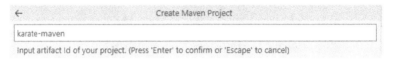

Figure 2.37 – Specifying the artifact ID

Now, you can select a folder where this project should be created. Note that this is the parent folder in which a new project folder with the name of your artifact ID (in this case, karate-maven) will be created automatically.

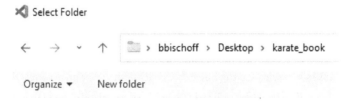

Figure 2.38 – Selecting the output folder

After selecting the folder, the assistant will continue in VS Code's terminal window. It will ask you what version you want to give your project. If you want to accept its suggestion, 1.0-SNAPSHOT, just press *Enter*:

```
[INFO] Using property: groupId = blog.softwaretester
[INFO] Using property: artifactId = karate-maven
Define value for property 'version' 1.0-SNAPSHOT: :
[INFO] Using property: package = blog.softwaretester
Confirm properties configuration:
groupId: blog.softwaretester
artifactId: karate-maven
version: 1.0-SNAPSHOT
```

```
package: blog.softwaretester
 Y: :
```

Now, Maven will present you with all your inputs again. If you want to make changes, enter the letter N (meaning *no*) and hit *Enter* one more time to restart the terminal prompts and enter new values. If you want to accept everything, just press *Enter* without entering any letter.

If you see BUILD SUCCESS, the project is created.

```
[INFO] -----------------------------------------------------
[INFO] BUILD SUCCESS
[INFO] -----------------------------------------------------
```

VS Code might prompt you to open this project right away, which is nice. This can be done anytime afterward through the **File | Open Folder** menu if you miss it. The project should appear in the explorer panel, as shown here:

Figure 2.39 – The opened Maven project

Included in the newly created project is the sample feature file under src/test/java/examples/ users/users.feature and the additional Karate runner classes, src/test/java/examples/ users/UsersRunner.java and src/test/java/examples/ExampleTest.java.

Also, two configuration files, src/test/java/karate-config.js and src/test/java/ logback-test.xml, are in here. These will play a role in the chapters to come.

Additionally, the Maven section in the explorer panel should be able to show the current dependencies. In our case, a version of com.intuit.karate:karate-junit5 should appear, since this is part of the Karate Maven archetype.

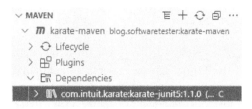

Figure 2.40 – The Maven dependencies view

A project without an archetype

The Karate archetype is a good start for a new project. Of course, it is also possible to start a blank Maven project and add the necessary dependencies and configurations manually if desired. To get started and check out what an ideal setup should look like, I recommend starting with the archetype and then modifying it afterward if needed.

In this section, we looked at how to install and configure Maven, as well as how to set up a new Karate project using the Maven Karate archetype. Next, let's look at some of the files that were generated for us by the Maven archetype.

The roles of the different Karate project files

The Maven archetype created some files for us that are important in this book. Let's go through them and check what they are used for, in the order they were generated in the project:

- `Users.feature`: This is an example test scenario of the archetype. Later, it can be removed when we add our own scenarios, but for now, it is good to keep it as a reference. Working with the different functionalities of feature files will be the main topic in *Chapter 3, Writing Basic Karate Tests*.

- **Test runner classes**:

  - `UsersRunner.java`: This is a sample runner class that showcases how to run a specific scenario. This will be discussed in more detail in *Chapter 4, Running Karate Tests*.

  - `ExamplesTest`: This one is a sample runner class that demonstrates how to run all tests in sequence or parallel. We will also discuss this in more detail in *Chapter 4, Running Karate Tests*.

- `karate-config.js`: This is Karate's central JavaScript configuration file. This is optional for simple tests where everything is specified within the feature files. Whenever we want to have central properties or JavaScript functions that should be reused across multiple scenarios, we will need this file for sure. This will be most important in *Chapter 6, More Advanced Karate Features*, and *Chapter 7, Extending Karate Functionality*.

- `logback-test.xml`: This is the main configuration file for Karate's logging. We will look at this one again in *Chapter 5, Reporting and Logging*.

- `pom.xml`: This is the main Maven configuration that includes dependencies to different libraries (such as Karate) and their version numbers, configuration information, and build plugins. For now, this is enough information. In *Chapter 4, Running Karate Tests*, we will look at this file in more detail.

In this section, we checked out what the different files in the Karate archetype project mean. These files will play a big role in the chapters to come.

Summary

In this chapter, we talked about the prerequisites for working effectively with Karate, namely installing Java, preparing an IDE, and how to use the Karate Maven archetype to create a new project. Additionally, we looked at the Karate standalone project, which lets us quickly explore the functionality of the framework.

In *Chapter 3, Writing Basic Karate Tests*, we will start using Karate's API testing functionality to write tests against a basic API.

You will learn how to call endpoints, use Karate's matches, and write efficient test scenarios without code duplication.

3
Writing Basic Karate Tests

Now that we have completed the system preparation and IDE setup in the last chapter, we can directly dive into the main topic: writing Karate API tests.

Apart from the basic Karate commands and structure, we will also take a closer look at how to make test scenarios easier to read and more straightforward. When you have dozens or hundreds of scenarios, it is especially beneficial to make them as readable and understandable as possible, so you don't have to think too much about what a scenario does and how the data looks.

In this chapter, we will cover these main topics:

- Exploring the API under test
- Calling endpoints and setting parameters
- Matching status codes and responses
- Making requests with payloads
- Using variables and data tables

Technical requirements

The code examples for this chapter can be found at `https://github.com/PacktPublishing/Writing-API-Tests-with-Karate/tree/main/chapter03`.

You will require the following:

- The system and IDE setup we completed in *Chapter 2, Setting Up Your Karate Project*.
- **Postman** (available via `https://www.postman.com/downloads`) for exploring the APIs we want to test later.
- Please note that when you launch Postman after the installation, you don't have to create a new account. Just skip this step by clicking on the highlighted link as follows, which is easy to overlook:

Figure 3.1 – Postman launch

Let's check out the API we will use in this chapter next!

Exploring the API under test

The first step in writing effective API tests is to explore the API under test. This can be done both by reading the API documentation and by playing around with it. This way, you can already get a feel for how the interfaces are structured and where there might be sources of errors that should be tested automatically.

The JSONPlaceholder API

We will use the *JSONPlaceholder* API for this chapter. This is a publicly available mock API that simulates a real API. Unlike a real REST application, the data here is not generated dynamically, but fixed static data is returned per endpoint. This makes it ideal for illustrating some Karate concepts. Plus, for now, we don't have to worry about dealing with mutable return values that are harder to check since they are more unpredictable.

Later in this chapter, we will see how we can still make checks against server-side responses without knowing exactly what values they include.

> **Mock APIs versus real APIs**
>
> While, in theory, it is possible to use any real REST API for this chapter, the examples in this chapter will mainly use the mentioned **JSONPlaceholder API**. Once you understand the main principles of Karate API tests, it should be straightforward to test any other REST API.

Checking the documentation

According to the JSONPlaceholder documentation at `https://jsonplaceholder.typicode.com`, we have the following resources available:

/posts	100 posts
/comments	500 comments
/albums	100 albums
/photos	5000 photos
/todos	200 todos
/users	10 users

Figure 3.2 – Available endpoints in JSONPlaceholder

This is the complete list of all data structures we can use with this API.

In this chapter, we will focus on the `/posts` endpoint. It simulates a list of blog posts, with each one having a title, body (the actual text of the post), and a user ID that links it to a specific user account.

We can see that each one returns a fixed list of entries, for example, the `/posts` resource always returns exactly *100* posts (given we don't use any filtering). In a real API, this list would be dynamic, of course.

The API documentation also gives us a list of *routes*, meaning the requests that we can make to each endpoint:

GET	/posts
GET	/posts/1
GET	/posts/1/comments
GET	/comments?postId=1
POST	/posts
PUT	/posts/1
PATCH	/posts/1
DELETE	/posts/1

Figure 3.3 – Available routes in JSONPlaceholder

While API documentation is generally a good way to start exploring, an even better one can be using the API yourself.

To check out the behavior and return values of the resources and routes, you can use a web browser, the command line, or a specialized REST API exploration tool. We will use **Postman** in the next step.

Using Postman

Postman is an ideal tool to interact with an API in order to gain a deeper understanding of it.

> **Proceeding without Postman**
>
> If you don't want to install or use Postman, you can also do basic API exploration using a web browser. If you paste an API URL into your browser, it will do a GET call and show you the data you receive.
>
> It is also possible to use a command-line tool, such as **cURL** (https://curl.se). However, a dedicated UI-based tool is generally easier and more comfortable to work with.

If you start Postman, you will see an empty collection. This is where you can group API requests that belong together:

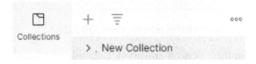

Figure 3.4 – Empty Postman collection

To rename this collection something more meaningful, right-click on it and choose **Rename**:

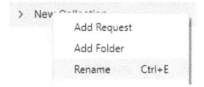

Figure 3.5 – Renaming a Postman collection

Let's call it JSONPlaceholder so we can gradually add more requests to it that we want to execute against this API:

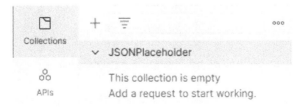

Figure 3.6 – Renamed Postman collection

Click on **Add a request** to specify the first GET request we want to check out. Choose **GET** as the method and `https://jsonplaceholder.typicode.com/posts` as the request URL. You can also give this request a more descriptive name. I chose Get all posts:

Figure 3.7 – Adding a new request in Postman

To trigger one of the saved requests, click **Send**. This will perform a GET request on the posts endpoint and return the JSON response so we can take a closer look at it:

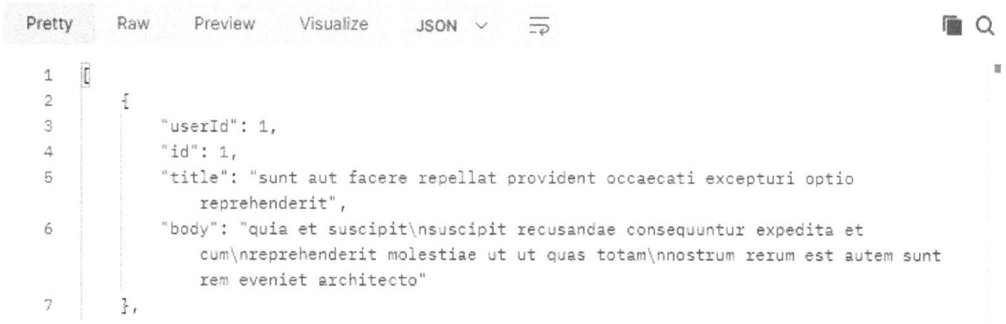

```
Pretty    Raw    Preview    Visualize    JSON  v

1    [
2        {
3            "userId": 1,
4            "id": 1,
5            "title": "sunt aut facere repellat provident occaecati excepturi optio
                 reprehenderit",
6            "body": "quia et suscipit\nsuscipit recusandae consequuntur expedita et
                 cum\nreprehenderit molestiae ut ut quas totam\nnostrum rerum est autem sunt
                 rem eveniet architecto"
7        },
```

Figure 3.8 – Sending a request and showing the response in Postman

According to the API documentation, we can retrieve posts belonging to a specific user by providing a userId parameter, like this:

```
https://jsonplaceholder.typicode.com/posts?userId=1
```

Here, userId=1 specifies that we only want to see posts with a user ID of 1. Let's add a second request to our Postman collection to store this request as well:

Figure 3.9 – Adding another request in Postman

Now, we can specify the complete URL, including the parameter. You will notice that Postman adds this parameter so we can directly modify or even disable it without needing to change the URL:

Figure 3.10 – Postman request with a parameter

Calling this new request in Postman shows us the expected result: a list of JSON posts that all have userId set to 1.

This will be the first endpoint that we will test using Karate in the next section.

Creating a new Karate project

Let's create a new Karate project by following the steps from *Chapter 2* in the *Setting up a Karate project with the Maven archetype* section.

I will call it karate-jsonplaceholder since this is a good description of what it will test. After this is done, you should have a project like this in VS Code:

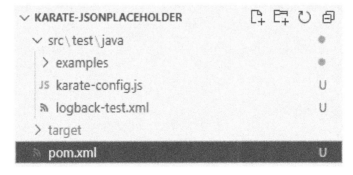

Figure 3.11 – New Karate project for JSONPlaceholder

Since we will write our tests ourselves and don't need additional configurations for now, we can delete the complete examples folder as well as the karate-config.js file. So now, your project should look like this, ready to be filled by us:

Figure 3.12 – Cleaned up Karate project for JSONPlaceholder

In the next step, we will create our first test case.

Adding a new feature file

Let's add a new scenarios folder under src/test/java that our feature files will be in:

Figure 3.13 – Creating a new folder

The next step is to add a new feature file inside that folder that will hold the scenarios belonging to the same business case:

Figure 3.14 – Created scenarios folder

We want to test the posts endpoint first, so let's create a new file:

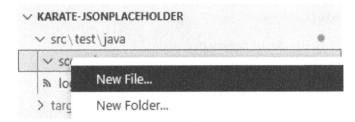

Figure 3.15 – Creating a new file

We will call it posts.feature since it deals with testing the posts endpoint. Notice that it should have a Karate icon to the left since the Karate plugin recognizes the .feature file extension:

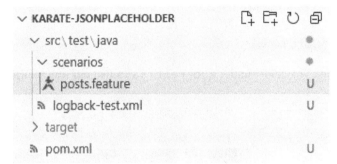

Figure 3.16 – posts.feature file with the Karate icon

Open the new posts.feature file and start by adding the mandatory Feature: keyword. Right after it, we will add a description of Posts endpoint tests to be extra clear:

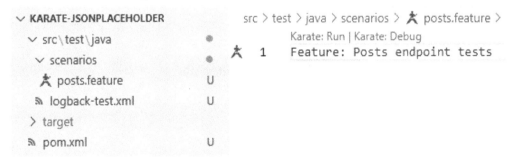

Figure 3.17 – New feature file

Now that we have a feature file, let's quickly configure the test runs from the IDE so we can easily check later that everything works.

Configuring test runs

When opening feature files, you will see a so-called **CodeLens**. This is a small display above some executable lines in the file that contains the two actions **Karate: Run** and **Karate: Debug**. For now, we will only use the run feature (debugging will be discussed in *Chapter 4, Running Karate Tests*).

Figure 3.18 – Karate CodeLens

If you click on **Karate: Run** now, you will see that another dialog opens that prompts for a Karate Runner file. We don't need this for now as this will be a topic for *Chapter 4*. Instead, let's configure the Karate plugin to use an alternative way to run the tests:

Figure 3.19 – Karate Runner prompt

Clicking on the Karate icon in the top right of VS Code and choosing **Open Settings** will go to the **Karate Runner** plugin settings page (alternatively, you can go there via the **File | Preferences | Settings** menu):

Figure 3.20 – Karate CLI activation

This will use the Karate command line to launch tests with Maven instead of requiring a dedicated runner file. When going back to the feature file and clicking **Karate: Run** now, you should see a lot of output in the terminal:

```
==========================================================
elapsed:   0.34 | threads:     1 | thread time: 0.00
features:     0 | skipped:     1 | efficiency: 0.00
scenarios:    0 | passed:      0 | failed: 0
==========================================================
```

It is correct that we get the report stating that no features and no scenarios were run since our feature does not contain any scenarios yet. That's why we also don't get a number for `passed`.

Adding a scenario

Let's add a scenario under the created feature that should test whether the posts endpoint is available:

```
Feature: Posts endpoint tests
    Scenario: Checking user specific posts
```

You will notice that the Karate plugin also adds a CodeLens above the scenario as well, making it possible to not only run everything in a feature file but also select which scenario to run.

Trying to click on **Karate: Run** above this scenario should give us a changed result now since we now have a scenario belonging to the feature:

```
==========================================================
elapsed:   3.70 | threads:     1 | thread time: 0.00
features:     1 | skipped:     0 | efficiency: 0.00
scenarios:    1 | passed:      1 | failed: 0
==========================================================
```

However, of course, we did not implement any steps yet. Therefore, our scenario is marked as `passed` – nothing is there that could fail yet. We will implement some steps in the next section.

Calling endpoints and setting parameters

In this section, we will learn more about calling common endpoints and setting parameters.

Setting a URL

Let's discuss the basic operation that every Karate API test starts with – calling a specific URL. This is achieved using the `url` keyword. Since this is our initial condition, we can use the `Given` keyword here:

```
Scenario: Checking user specific posts
    Given url 'https://jsonplaceholder.typicode.com/
posts?userId=1'
```

It is important to put the URL in quotes – otherwise, you will get an error. This is due to how Karate parses the steps using regular expressions, as we saw in *Chapter 1*. Also, it must be in the same line as `Given url`; it just did not fit here due to the limitation in line length.

> **Endpoint URL specification**
>
> In this first example, the endpoint URL is the first line in our scenario. This is more for demonstration purposes, as in the real world, a test is more likely to run against different environments and base URLs. Also, there is a high chance that multiple scenarios use the same base URLs. Later in this book, we will go over different ways to centralize the URL configuration.

This line does not call the URL yet but only sets it up for the subsequent steps. The set URL will be active until the next `url` keyword in a scenario is used. This can be confusing at first, but it makes sense, especially when we look at how to set paths and parameters separately later.

Specifying the HTTP method

Next, we need to specify an HTTP method (GET, POST, PUT, DELETE, PATCH, and so on) that we want to use with the specified endpoint. In our case, we want to use GET as this is a simple call to retrieve all posts. Fortunately, Karate's convention uses the exact same names for these methods (also called **verbs**) as keywords, just in lowercase, preceded by `method`:

```
Scenario: Checking user specific of posts
    Given url 'https://jsonplaceholder.typicode.com/
posts?userId=1'
        When method get
```

We use `When` for this step since this can be seen as an *action* performed on the currently set URL.

Running the test now again with **Karate: Run** in the scenario's CodeLens should give us an output like before because we did not specify any assertions about what we expect to happen when calling this URL. So far, it is not a test yet but just a chain of commands to call a URL.

What we should see, though, is the complete request in the logs when scrolling up, including the method, URL, and some headers:

```
16:47:25.928 [com.intuit.karate.cli.Main.main()] DEBUG com.
intuit.karate - request:
1 > GET https://jsonplaceholder.typicode.com/posts?userId=1
1 > Host: jsonplaceholder.typicode.com
1 > Connection: Keep-Alive
1 > User-Agent: Apache-HttpClient/4.5.13 (Java/17.0.4.1)
1 > Accept-Encoding: gzip,deflate
```

Also, Karate will show the complete response in the logs too, including the time it took to receive the response, the status code, the date and time, the content type, and various other pieces of information:

```
16:47:26.380 [com.intuit.karate.cli.Main.main()] DEBUG com.
intuit.karate - response time in milliseconds: 437
1 < 200
1 < Date: Wed, 12 Oct 2022 14:47:26 GMT
1 < Content-Type: application/json; charset=utf-8
1 < Transfer-Encoding: chunked
1 < Connection: keep-alive
1 < X-Powered-By: Express
1 < X-Ratelimit-Limit: 1000
```

After this block, Karate even prints the whole JSON data we receive back as well by default (only the first set of data is shown here):

```
[
  {
    "userId": 1,
    "id": 1,
    "title": "sunt aut facere repellat provident occaecati
excepturi optio reprehenderit",
    "body": "quia et suscipit\nsuscipit recusandae consequuntur
expedita et
cum\nreprehenderit molestiae ut quas totam\nnostrum rerum est
autem sunt rem
eveniet architecto"
```

```
    },
    ...
    ]
```

Let's find out how we can make this test more readable next.

Separating the base URL from the path

You might have noticed that we currently specify the complete endpoint URL in the first step:

1. The base URL, `https://jsonplaceholder.typicode.com`, is the common URL for all endpoints.

2. The `/posts` part is the concrete resource path in this case.

Considering the later maintenance and reusability of the code, it is better to separate these two so we will be able to set the base URL from a different place in the future:

```
Scenario: Checking user specific posts
    Given url 'https://jsonplaceholder.typicode.com'
    And path 'posts'
    When method get
```

Using the `path` keyword adds the specified path to the base URL. Note that you don't need to add a slash – this is done by Karate automatically for us.

This manipulation of the URL to call is possible since the URL is only set here and not called yet. We can even see in the logs that the URL including the path is called, almost as we had before:

```
1 > GET https://jsonplaceholder.typicode.com/posts
```

Now, we still need to take care of the `userId` query parameter.

Setting query parameters

Lastly, let's set the query parameters.

```
Scenario: Checking user specific posts
    Given url 'https://jsonplaceholder.typicode.com'
    And path 'posts'
    And param userId = 1
```

Note the spaces before and after = as this is important for Karate to recognize it. Like the path step, we use the `And` keyword because it is connected to the first `Given` step.

We don't need to add a question mark or ampersand – Karate constructs the correct URL automatically.

When running the test now, we can see in the logs that the correct URL, including the query parameter, is requested:

```
13:06:32.602 [com.intuit.karate.cli.Main.main()] DEBUG com.
intuit.karate - request:
1 > GET https://jsonplaceholder.typicode.com/posts?userId=1
```

Also, we should only get JSON elements back with `"userId": 1` as this is the `userId` value we requested.

We now have a good starting point for the next section, where we will check for the status code and response type as well as run checks on the returned JSON response to turn our first scenario into a real test case.

Matching status codes and responses

In the last output, we saw that we get a status code of 200 (also known as *OK*). This means the request was successful. Let's add a step to verify this.

Matching the status code and type

If you want to match a simple status code, this is done via the `status` keyword like this:

```
Scenario: Checking user specific posts
    Given url 'https://jsonplaceholder.typicode.com/posts'
    When method get
    Then status 200
```

We use the Then keyword here to indicate that this is an assertion. At this point, we can run the test (which should pass).

> **Status codes**
>
> For a complete list of HTTP status codes, see `https://developer.mozilla.org/en-US/docs/Web/HTTP/Status`.

It is a good idea to deliberately make it fail once to verify that this test would not give us a false positive result in an error case.

Making the test fail

Let's change the status code to the wrong one and run the test again:

```
Then status 404
```

As expected, the test fails and even gives us the error reason in the log:

```
============================================================
elapsed:    4.68 | threads:      1 | thread time: 1.38
features:      1 | skipped:       0 | efficiency: 0.30
scenarios:     1 | passed:        0 | failed: 1
============================================================
>>> failed features:
status code was: 200, expected: 404, response time in
milliseconds: 478
```

Additionally, in VS Code, we can see the number of passed and failed tests as well as the overall failure status in the IDE footer:

⊗ 10/14/2022 14:48:51.526

c:\Users\bbischoff\Desktop\github...st\java\scenarios\posts.feature:2

Features: 1 | Scenarios: 1 | Passed: 0 | Failed: 1 | Elapsed: 4.33ms

↷ ☰ ⚙ ⊕ 🔖

Karate ⊘ 0 ⊗ 1   ⊗ 0 △ 2

Figure 3.21 – Failed test run in red in VS Code

In the code view, this will also be indicated by the failed step being underlined. This will go away once the test passes.

Let's change the response code to 200 again to make the test pass:

Figure 3.22 – Passed test in VS code

We can see this in the VS Code footer due to the number **1** beside the check icon. Also, the background color is not red anymore.

Let's add some more checks for the actual JSON response!

Using assertions and matchers on the response

So far, we are not really doing any testing concerning the returned data but only the status code. To fulfill the title of our scenario (*Checking user-specific posts*), we must add at least one other step with an assertion so this can be a real test.

> **JSON examples**
>
> All examples in this chapter apply to JSON data since this is the most common REST API request-and-response format. For XML-based APIs, the shown approaches are almost identical. More on working with matchers on XML data can be found here: `https://github.com/ karatelabs/karate#xpath-functions`.

From our API exploration and its documentation, we know that the endpoint in use returns a list of posts where each one has a `userId` that should be equal to the one that we pass via the request parameter. Additionally, it has an `id` that can identify each post, `title`, and `body`. The following is an example structure of a posts resource:

```
{
    "userId": 1,
    "id": 1,
    "title": "sunt aut facere repellat provident occaecati
excepturi optio reprehenderit",
    "body": "quia et suscipit\nsuscipit recusandae consequuntur
expedita et cum\nreprehenderit molestiae ut quas totam\nnostrum
rerum est autem sunt rem eveniet architecto"
}
```

Let's first check that we get at least one result back.

Exploring the response variable

Whenever any request is made, you will have the `response` variable available afterward. This one stores the complete response body from the last request. In our case, this will be the complete JSON response.

If we add a step to print the response, we will log our whole JSON body, as seen in the section before:

```
Scenario: Checking user specific posts
    Given url 'https://jsonplaceholder.typicode.com/posts'
    When method get
```

```
Then status 200
And match responseType == 'json'
* print "RESPONSE:", response
```

This step starts with a * character since it is not part of the usual scenario flow but only a debugging step without any other purpose.

The nice part of the response variable is that it includes the response body in the right format. That means that it is not just a big string that looks like JSON, it *is* JSON that we can directly work with.

We can directly verify this by changing the print statement to give us the first element of the response:

```
* print "RESPONSE:", response[0]
```

This one gives us just the following response:

```
23:13:11.660 [com.intuit.karate.cli.Main.main()] INFO  com.
intuit.karate - [print] RESPONSE: {
  "userId": 1,
  "id": 1,
  "title": "sunt aut facere repellat provident occaecati
excepturi optio reprehenderit",
  "body": "quia et suscipit\nsuscipit recusandae consequuntur
expedita et cum\nreprehenderit molestiae ut quas totam\nnostrum
rerum est autem sunt rem eveniet architecto"
}
```

The fact that we only have curly braces, { }, surrounding our value shows clearly that we only have *a single JSON object*. Otherwise, we would get an *array of elements* back, which is indicated by square brackets that contain a list of curly brace-surrounded objects, essentially like this: [{ }, { }, { }].

If we want to check for specific keys inside an element (for example, userId), we can directly access them with a dot, like this:

```
* print "First user id:", response[0].userId
```

This returns us only the userId of the first element, which is our expected value:

```
23:37:51.741 [com.intuit.karate.cli.Main.main()] INFO  com.
intuit.karate - [print] First user id: 1
```

This simple example shows the underlying power of Karate when it comes to JSON.

> **Shortcuts**
>
> Karate has many shortcuts that can drastically shorten the needed code even more. In the case of response, this can also be written as $. This is often used in more complex expressions, as we will explore further in this chapter. In this book, I will use both ways interchangeably, depending on the length of the code.

Let's take a quick look at how nested JSON elements can be accessed.

Handling nested JSON elements

This also works with nested JSON elements within other elements. Let's take the JSON example from *Chapter 1*:

```
{
    "book": {
        "title": "Writing tests with Karate",
        "author": "Benjamin Bischoff",
        "chapters": [{
            "number": "1",
            "title": "Karate's core concepts"
        }]
    }
}
```

In this case, we could use a construct such as response.book.title, which would give us the title of the book. If we want the title of the first chapter, we would have to use response.book.chapters[0].title.

Let's get back to our current example and do some more checks.

Asserting the number of elements with the length property

We know that we get some elements back from our posts API if we pass in a user ID. So, in this step, we will assert this.

Let's first print out the number of elements within our response using the length property of the response:

```
* print "Length", response.length
```

The `length` property can be used on any array within the response. In our case, the response is already an array of posts, so we can directly use it here. When running the test, it returns this:

```
21:21:38.148 [com.intuit.karate.cli.Main.main()] INFO  com.
intuit.karate - [print] Length 10
```

Karate has a powerful keyword called `assert`, which can be used for any expressions that evaluate to `true` or `false`. This makes it well suited for our example since we want to assert whether we get at least one post resource for *userId 1*.

So, let's add this step to our scenario:

```
And assert response.length > 1
```

This checks whether the number of elements in the response array is greater than 1. The scenario should still pass at this point.

Let's temporarily change this step to a condition that should fail, like this:

```
And assert response.length > 1000
```

If we run it now, the test fails as expected and the Karate logs tell us clearly what happened. It even points out the condition, `response.length > 1000`, as the failure cause:

```
>>> failed features:
did not evaluate to 'true': response.length > 1000
src/test/java/scenarios/posts.feature:11
```

This is good already because we know that the test failed and why. We would not know what the current length of the array is, though, unless we deliberately print it out. In the real world, this would lead to three steps:

1. See that the test is failing.
2. Add a `print` statement to find out exactly why it fails.
3. Rerun the test to check the value of the `print` statement.

There is another way of doing this, which can give us the desired information right away. Let's look at **matchers** in the next section.

Using matchers

As we have seen, `assert` can tell us whether something is true. Matchers, on the other hand, are there to compare expected properties and values of elements (or variables, as we will see later) and give us information about the actual state if something fails.

All matcher steps use the `match` keyword followed by either an expression or another keyword specifying how the matcher should perform its checks. This will become clearer in the following examples where we will explore the power of the different matcher options.

Checking the response type

If you want to verify that a response is of a certain format, this can also be done easily. Let's add another check to see that we get the correct response type (in our case, JSON) back from the API call. For this, we can use the `match` and `responseType` keywords together:

```
Scenario: Checking user specific posts
    Given url 'https://jsonplaceholder.typicode.com/posts'
    When method get
    Then status 200
    And match responseType == 'json'
```

`responseType` can be set to `json`, `string`, or `xml`. Please be aware of the double equal signs (`==`) for value comparison. Also note that we use the And keyword here because it is an additional check after the status code verification, so basically, another Then step.

Let's look at the next use case of `match`, which is checking for specific values.

Matching values

The simplest matching case is if we want to check whether an element has a specific value. Say we want to verify that `userId` is really set to 1 in the first returned post. This can be done by using ==, as we have seen before in the `responseType` example:

```
And match response[0].userId == 1
```

In contrast to doing this with `assert`, this is more helpful when it fails. In a failure case, for example, if `userId` was 2, Karate would give us this message:

```
And match response[0].userId == 1
match failed: EQUALS
  $ | not equal (NUMBER:NUMBER)
  2
  1
```

This is much clearer than before! Of course, you can also use ! = if you want to check that a value is not equal to the given one:

```
And match response[0].userId != 1000
```

In our example, we just assumed that the first element has a `userId` key. In a real-world test, we should not do this but, instead, check first that our expected keys really exist before asserting specific values. We will do this next.

Checking the response elements' existence with match contains

To check whether a response contains some expected keys and values at the same time, we can use `match contains`. Our obvious need is to check whether `userId` is 1, so this is a good way to do it without having to worry about specifying all other keys in the response as well:

```
And match response[0] contains { userId: 1 }
```

If you want to check whether a specific element is *not* included in the response, you can do that using `!contains`:

```
And match response[0] !contains { unknownElement: 'test' }
```

In both cases, you are, of course, not limited to single values, but you can also match multiple key-value pairs:

```
And match response[0] contains { userId: 1, id: 1 }
```

Lastly, if there are multiple possibilities of values that you want to check, you can also use `match contains any`, meaning it will check whether one or more of your elements matches:

```
And match response[0] contains { userId: 1, userId: 2 }
```

This would also be true in our case because the first response does contain `userId: 1`. However, this would also be successful if we had `usedId: 2`, or even both!

For more specialized matchers, please check out the official documentation: `https://github.com/karatelabs/karate#matching-sub-sets-of-json-keys-and-arrays`.

Let's see how fuzzy matchers can help us next.

Using fuzzy matchers

Fuzzy matchers are used when you don't know the exact values but the structure or data type of the elements. Also, these are handy if you don't want to match a whole element but only some specific keys and values.

Here is a reference element again:

```
{
    "userId": 1,
    "id": 1,
```

```
    "title": "sunt aut facere repellat provident occaecati
excepturi optio reprehenderit",
    "body": "quilaxersaz
a et suscipit\nsuscipit recusandae consequuntur expedita et
cum\nreprehenderit molestiae ut quas totam\nnostrum rerum est
autem sunt rem eveniet architecto"
}
```

Let's check whether all keys that should be there in a post element (userId, id, title, and body) exist, regardless of their values. This can be achieved by constructing a JSON structure that has the same keys as our source and special markers (starting with a # symbol) that can check for different things.

Since we want to check first whether all keys are there, we can use the #present marker:

```
And match response[0] == { id: '#present', userId: '#present',
title: '#present', body: '#present' }
```

This is an *exact* match of the schema. In this case, if the response had an additional key that was not mentioned in the match expression, it would fail.

If you want to check the opposite, so whether an element should *not* exist in a specific case, you can use the #notpresent marker instead:

```
And match response[0] == { unknownElement: '#notpresent',
id: '#present', userId: '#present', title: '#present', body:
'#present' }
```

This test would still be successful since an element named unknownElement is indeed *not* included in our response.

Apart from #present and #notpresent, there are many more of these markers that check different things. For example, you can use #null and #notnull to determine whether the element has a null value or doesn't, respectively.

Instead of verifying whether an element is there or not or has a value or not, you can also assert the specific data type. Just use #array, #object, #boolean, #number, or #string to achieve this.

Let's look at an example for our use case:

```
And match response[0] == { id: '#number', userId: '#number',
title: '#string', body: '#string' }
```

This is successful – all the data types match our expectations: id and userId are both *numbers*, whereas title and body are *string values*.

Again, if we change one to a wrong data type (for example, `id: '#string'`), we get a very thorough error message like this:

```
match failed: EQUALS
  $ | not equal | match failed for name: 'id' (MAP:MAP)
  {"userId":1,"id":1,"title":"sunt aut facere repellat
provident occaecati excepturi optio reprehenderit","body":"quia
et suscipit\nsuscipit recusandae consequuntur expedita et cum\
nreprehenderit molestiae ut quas totam\nnostrum rerum est autem
sunt rem eveniet architecto"}
{"userId":"#number","id":"#string","title":"#string",
"body":"#string"}
    $.id | not a string (NUMBER:STRING)
    1
    '#string'
```

This does not only tell us that the type check failed for the `id` field but also gives us *the exact matcher string* we used and *the full JSON element that caused the failure*! This makes finding the cause of bugs straightforward.

> **Optional values**
>
> While you can use #null or #notnull, or #present or #notpresent, to check whether an element has a specific condition, you can also use another way to check for optional values.
>
> You can add another # to the marker to make it an optional check (e.g., { 'userId': '##number', 'id': '##string', … }). This means that these values could be missing or set to null. But *if* they are there, they will be checked against the specified data type. This makes these conditions very flexible.
>
> If you do *not* want to check whether a specific element exists or not, you can use the #ignore marker as well.

So far, we have only checked one array element: `response[0]`. Let's check out how we can do this for all elements.

Matching all array elements

To go through all array elements and check their data respectively, we can use a special element to look at a specific form of matchers: `match each`.

We can use a statement like this to validate the data types of each element:

```
And match each response == { id: '#number', userId: '#number',
title: '#string', body: '#string' }
```

Note that we don't have to use `response[0]` in this case but just the full `response` array without an index. `each` will go through all array elements automatically.

It is also possible to go through the `each` elements in conjunction with `contains`. This makes it possible to verify that all entries we get back contain the correct `userId`, like so:

```
And match each response contains { userId: 1 }
```

This is very helpful because it not only tells Karate what went wrong but also in which specific element index. This is followed by a complete printout of the element in question and is finally completed by the error:

```
    $[0] | actual does not contain expected | all key-
values did not match, expected has un-matched keys - [userId]
(MAP:MAP)
    {"userId":2,"id":1,"title":"sunt aut facere repellat
provident occaecati excepturi optio reprehenderit","body":"quia
et suscipit\nsuscipit recusandae consequuntur expedita et cum\
nreprehenderit molestiae ut quas totam\nnostrum rerum est autem
sunt rem eveniet architecto"}
    {"userId":1}
      $[0].userId | not equal (NUMBER:NUMBER)
      2
      1
```

In this case, this means that the first array element contained `userId` 2 where 1 was expected. The `match each` command fails and stops checking other elements as soon as there is an entry that does not match.

In the next section, we will see how we can validate the schema of our JSON structure.

Validating the JSON structure

Typically, validating JSON structures is done by using **JSON Schema**, a specific set of instructions that can be used to describe and validate JSON data. However, this can be rather complex to use (see `http://json-schema.org/learn/getting-started-step-by-step.html`).

Karate has its own way of doing this in a much easier way.

Validating arrays

There is a *specific marker for arrays* that can be used in conjunction with `match` to do more elaborate array validations. This looks like this:

```
And match response == '#[]'
```

This validates that the given `response` value should be an array.

The nice thing about this syntax is that we can also use this to check for a specific size of arrays if we just enter the size in between the square brackets:

```
And match response == '#[10]'
```

This means that we expect `response` to be an array with *10* entries. For our test API, this is the case. If it weren't (let's say we get an array with *9* elements back instead), we would get another helpful error message like this:

```
match failed: EQUALS
  $ | actual array length is 9 (LIST:STRING)
```

We can even go further and tell the array what data type we expect it to have, as seen before. For this, we can add it after the array marker:

```
And match response == '#[] #object'
```

Now, it gets interesting, since we can do multiple checks at the same time.

This line might look complex, but it is straightforward once you get to know the syntax better:

```
And match response == '#[]? _.userId == 1'
```

This line combines two checks into one:

1. `#[]`: Is our response an array?
2. `? _.userId == 1`: Does each array element have a `userId` property with a value of 1?

The `?` symbol signals that we want to iterate over the array, whereas the `_` symbol specifies each element that we are iterating over.

With this technique, it is even possible to combine even more logic into one single line:

```
And match response == '#[10] #object? _.userId == 1'
```

This example includes four different checks:

1. `#[]`: Is our response an array?
2. `#[10]`: Does the array have 10 elements?
3. `#object`: Does our array contain objects?
4. `? _.userId == 1`: Does each array element have a `userId` property of 1?

There are even more ways to use this powerful feature – for example, for very specific checks against custom business rules using functions. We will return to this later in *Chapter 6, More Advanced Karate Features*.

For now, let's check out how to make requests with JSON payloads.

Making requests with payloads

So far, we have only worked with the responses. But what about making requests that include payloads such as POST, PUT, and PATCH? Let's look at another part of the JSONPlaceholder API that allows us to do just that.

> **A non-persistent mock API**
>
> It is important to know that the JSONPlaceholder API is non-persistent. That means that its underlying data is not changed regardless of our requests to create or modify data. It merely simulates that new data is created by delivering the appropriate response as a real API would. This makes it perfect for testing.

In the user guide, we can see that the /posts endpoint supports the POST operation to create a new post for a specific user.

According to the documentation, a request body to send to the endpoint should look like this:

```
{
    title: 'foo',
    body: 'bar',
    userId: 1,
}
```

The return value is supposed to be the created post including its ID:

```
{
  id: 101,
  title: 'foo',
  body: 'bar',
  userId: 1
}
```

Since the request body and response are simple in this case, it is a good example to illustrate post requests. Also, we can combine sending payloads and checking for the same value in the received response.

Let's start with a new scenario in our `posts.feature` file called `Creating a new post`, with the following content:

```
Scenario: Creating a new post
    Given url 'https://jsonplaceholder.typicode.com'
    And path 'posts'
```

You will notice that the first two lines are the same as the previous scenario because we also use the `posts` endpoint. This time, we just change what we do with it. Remember that, at this point, we are not calling the endpoint yet!

Now we need the payload to create a new post. Let's pretend we want to have a new post for user ID `10` with a title of `Hello` and `World` as body text. Since Karate is specialized in JSON data, we can just pass it as is using the `request` keyword:

```
And request { userId: 10, title: 'Hello', body: 'World' }
```

The last thing to use is to finish with the method as we did with `get` before – but this time using `post`, of course:

```
When method post
```

Notice I am using `When` here, as there will be more `Then` steps with some checks coming soon.

When running the scenario now (as before, by clicking on the **Karate: run** CodeLens link above the new scenario), this should pass and return the newly created post JSON in the logs:

```
{
  "userId": 10,
  "title": "Hello",
  "body": "World",
  "id": 101
}
```

In this case, we want to check whether our original request is included in the response, plus an additional ID.

So, we could do something along these lines:

```
Scenario: Creating a new post
    Given url 'https://jsonplaceholder.typicode.com'
    And path 'posts'
    And request { userId: 10, title: 'Hello', body: 'World' }
```

```
    When method post
    Then status 201
    And match responseType == 'json'
    And match response == { id: #number, userId: 10, title:
'Hello', body: 'World' }
```

Note, that our API under test returns a status code of 201 and not 200 in this case. 201 means that a resource was successfully created on the server by our POST request, whereas 200 tells us that a request was successful, but no new resource was created.

This works fine now. However, there is quite some duplication here because we check for almost the same JSON that we posted. The only difference is that we added the id: #number element to check for the presence of an id of the number type in the returned response.

So, in the next section, we will use Karate variables and data tables to reuse some code and thus make it clearer and easier to maintain in the process.

Using variables and data tables

Karate allows saving values into variables so we can use the same JSON bodies or values multiple times within one test and don't have to change them in various places if we want to test with different values. Data tables, on the other hand, allow for a more readable definition of JSON data.

Let's look at how we can use a variable to reduce the amount of code from our previous scenario.

Using variables

Variables are a very powerful tool in Karate tests. They can hold basically any value type, so they are very flexible.

To declare a simple variable, you can use the def keyword, meaning *define*. In the following case example, we store the Benjamin string in the myName variable and use it in the print statement:

```
Scenario: Declaring a variable
    * def myName = 'Benjamin'
    * print 'Hello from', myName
```

When we run it, this logs Hello from Benjamin, showing us the contents of myName as expected:

```
16:58:58.119 [com.intuit.karate.cli.Main.main()] INFO  com.
intuit.karate - [print] Hello from Benjamin
```

Be careful with this because *you can overwrite already existing variables*, which may lead to test failures down the line. If you keep your tests well designed and independent of each other, this will not happen, though.

> **Reserved keywords**
>
> Karate's `url` and `request` keywords are reserved keywords that may not be used as variables. According to the official documentation, this is to lessen confusion for the test author.

Variables can also hold JSON values (or even arrays of nested JSON), which makes them perfect for our current test case.

Let's first store our initial payload in a new variable called `payload` and use it to make our POST request as before:

```
Scenario: Creating a new post with variable
    Given url 'https://jsonplaceholder.typicode.com'
    And path 'posts'
    * def payload = { userId: 10, title: 'Hello', body: 'World'
}
    And request payload
```

I used the * symbol for the step that defines the `payload` variable since this is not a condition, action, or check but is just there to prepare some test data.

Notice that we can use the created `payload` variable seamlessly in our `request` step without any additional effort! This is yet another example of why the native JSON capability of Karate is so helpful.

Now we can use the same variable in our checks without repeating the request JSON one more time!

```
And match response == payload
```

This line tries to match our original `payload` variable to the received response. You could imagine that this works – but at this point, it does not yet. We can see the problem clearly when we run the test and look at the logs:

```
match failed: EQUALS
  $ | not equal | actual has 1 more key(s) than expected -
{"id":101} (MAP:MAP)
  {"userId":10,"title":"Hello","text":"World","id":101}
  {"userId":10,"title":"Hello","text":"World"}
```

The problem at this point is that our response is *not the exact request* that we passed but has an additional id variable. Now, at this point, there are multiple ways to solve this. I will show you probably the simplest solution next.

Our payload variable holds a JSON object. So, fortunately, we can just add the additional id key to our JSON variable before doing the check. And since we know, as before, that this should be a *number*, we can use the special #number marker here as well. Also, let's print out the modified payload variable so we can see that it works as expected:

```
* payload.id = "#number"
* print payload
And match response == payload
```

Now when the test is run, it should pass and print the new payload variable:

```
18:47:48.478 [com.intuit.karate.cli.Main.main()] INFO   com.
intuit.karate - [print] {
  "userId": 10,
  "title": "Hello",
  "text": "World",
  "id": "#number"
}
```

Wonderful, this works like a charm now.

For our current use case, the JSON request is not too complex. If you have an API that requires way more parameters in the payload, it can get a little bit hard to read. In this case, Karate offers an alternative way of specifying JSON: data tables!

Using data tables

Data tables are a way to define data for steps in a tabular way, which makes it clearer when dealing with complex JSON arrays.

Here is a basic example:

```
* table numbersAndWords
  | number | word   |
  | 5      | 'five' |
  | 10     | 'ten'  |
* print numbersAndWords
```

This prints out the following structure:

```
13:31:39.879 [com.intuit.karate.cli.Main.main()] INFO  com.
intuit.karate - [print] [
  {
    "number": 5,
    "word": "five"
  },
  {
    "number": 10,
    "word": "ten"
  }
]
```

This is how the data table is turned into JSON:

- The * table numbersAndWords line signals to Karate that a data table is expected that should be saved in the numbersAndWords variable.

- The first table line is the header; basically, each column acts as a key of the resulting JSON (number and word, in our case).

- Each line after the header is turned into a *JSON array element* with the keys from the headers and the values in this row.

Our test case would look like this with a data table:

```
Scenario: Creating a new post with a data table
    Given url 'https://jsonplaceholder.typicode.com'
    And path 'posts'
    * table payload
        |userId|title  |text   |
        |10    |'Hello'|'World'|
    And request payload[0]
    When method post
    Then status 201
    And match responseType == 'json'
    * payload[0].id = "#number"
    And match response == payload[0]
```

The downside of this approach is that we only have a single element, so we must add an index of 0 everywhere we use `payload` (for example, `payload[0]`). This is necessary because a data table is an array with each line being one array element. If you write tests with JSON arrays containing multiple elements, this is a great way to define them.

We can circumvent this if we just add a new variable like this:

```
Scenario: Creating a new post with a data table
    Given url 'https://jsonplaceholder.typicode.com'
    And path 'posts'
    * table payload
        |userId|title  |text   |
        |10    |'Hello'|'World'|
    * def payload = payload[0]
    And request payload
    When method post
    Then status 201
    And match responseType == 'json'
    * payload.id = "#number"
    And match response == payload
```

Here, we set the `payload` variable to the first element (`[0]`) of itself so we can continue using it as before.

Nested arrays

You can also write JSON into table cells, not only single values, to create even more complex array structures!

For elements containing many keys, these tables can get too wide. Here, we can use a set, which is basically a tipped-over data table.

Using a set

This is our test case using a set that is started with `set` instead of `table`:

```
Scenario: Creating a new post with a set
    Given url 'https://jsonplaceholder.typicode.com'
        And path 'posts'
        * set payload
            |path  | 0     |
```

```
        |userId|  10    |
        |title |'Hello'|
        |text  |'World'|
    * def payload = payload[0]
    And request payload
    When method post
    Then status 201
    And match responseType == 'json'
    * payload.id = "#number"
    And match response == payload
```

Here, the keys are in the left column and the values are in the right one. It is important to note that each subsequent column would be a new array element as well. Also, it is necessary to start the table with a header containing path for the key column and the array index above each value column.

I used the same trick of setting the resulting payload variable to its first element afterward to keep the following code clearer.

Omitting the index above the value column results in a single element being created so we can directly use it without any hassle!

```
Scenario: Creating a new post with a set
    Given url 'https://jsonplaceholder.typicode.com'
        And path 'posts'
        * set payload
            |path  |       |
            |userId|  10   |
            |title |'Hello'|
            |text  |'World'|
        And request payload
        When method post
        Then status 201
        And match responseType == 'json'
        * payload.id = "#number"
        And match response == payload
```

Summary

In this chapter, we talked about the basics of writing Karate API tests using different HTTP verbs and checks for status codes, content types, and various response matchers. Also, we checked out some ways to define payloads leading to a better understanding of our test scenarios.

In *Chapter 4*, *Running Karate Tests*, we will look at the various ways to run Karate tests, both from the IDE as well as from Maven. Also, we will check out some of the very helpful features of the Karate Runner VS Code plugin, which can be used to run and debug tests.

4

Running Karate Tests

Now that we know the basics of writing new Karate tests, we will explore different ways to run them through both VS Code and Maven. We will look at how to choose a specific environment to run tests against. We will also get to know how to run only a subset of tests, as well as discovering ways to efficiently debug them to pinpoint and fix issues more quickly.

In this chapter, we will cover these main topics:

- Running and debugging Karate tests through the IDE
- Running tests with Maven
- Running tests against different environments
- Running specific tests
- Filtering tests by tags

Technical requirements

The code examples for this chapter can be found at `https://github.com/PacktPublishing/Writing-API-Tests-with-Karate/tree/main/chapter04`.

You will also require the system and IDE setup we completed in *Chapter 2, Setting Up Your Karate Project*.

> **Example project**
>
> For this chapter, we will use an example project called `running-tests`, which is created using the standard Karate Maven archetype as is. As opposed to before, this time, we will use all generated files as they demonstrate the different run possibilities well.

Let's first check out some different ways we can run Karate tests straight from VS Code!

Running and debugging Karate tests through the IDE

When developing tests, it is vital that we can run them on our local development system. It does not stop there, though. Another very important part is that we have the tools available to properly debug the tests when developing them or do further exploration if something fails.

Running via CodeLens and the Karate CLI

In the last chapter, we already used the Karate plugin CodeLens to run specific tests by clicking on **Karate: Run**. Since this is a new project without any specific VS Code configuration, trying to trigger this will open this dialog to specify a Karate runner.

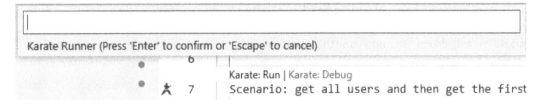

Figure 4.1 – Selecting a runner

We will look at runners later in this chapter. In the last chapter, we did not use a runner and instead chose the **Override Karate Runner** option as seen here.

Figure 4.2 – Overriding the Karate Runner

If this option is active, we can see that when we run a test through CodeLens, it is using the Karate `com.intuit.karate.cli.Main` command-line class and passing the test to run to it as an argument:

```
Executing task: mvn clean test-compile -f "c:\Users\bbischoff\
Desktop\github\Writing-API-Tests-with-Karate\chapter04\
running-tests\pom.xml" exec:java -Dexec.mainClass="com.intuit.
karate.cli.Main" -Dexec.args="c:\Users\bbischoff\Desktop\
```

```
github\Writing-API-Tests-with-Karate\chapter04\running-
tests\src\test\java\examples\users\users.feature:7" -Dexec.
classpathScope="test"
```

This was fine for our use case of simple triggering tests before. However, we won't use this here since we want to have more control.

More information about the CLI capabilities can be found at `https://github.com/karatelabs/karate/wiki/Debug-Server#karate-cli`. Let's look at how we can debug tests right from the IDE.

Debugging via CodeLens and Karate standalone

You might remember that we talked about Karate standalone in *Chapter 2, Setting Up Your Karate Project*, when we explored its basic functionality. In this section, we will use it again as a means for comfortably debugging our tests.

If we try to use the **Karate: Debug** option from CodeLens right now, it would not work. This is because there are two vital parts missing: the path to our instance of Karate standalone and a launch configuration to use it with.

Next, we will fix this.

Configuring Karate standalone

In this step, we will configure the `Karate standalone` path so it can be used for debugging. Open the settings by clicking on the upper-right **Karate** button and **Open Settings** (or via the **File | Preferences | Settings** menu).

Figure 4.3 – Open Settings

Look for the **Karate Runner > Karate Jar: Command Line Args** entry and replace `karate.jar` with the full path to `karate.jar` inside your Karate standalone directory.

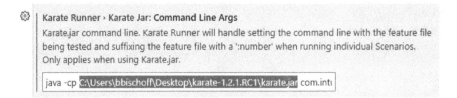

Figure 4.4 – Configuring the Karate JAR path

As you can see in the screenshot, ours is C:\Users\bbischoff\Desktop\karate-1.2.1.RC1\ karate.jar. Yours will of course be in another location and might have a different version number. It is important to leave the rest of the command intact.

> **VS Code settings**
>
> In case you are not aware of this, all the settings that we change are stored in a settings. json file under our project's .vscode directory. This makes it possible to make changes directly in there without having to go to the dedicated settings dialog.

The full command for **Runner > Karate Jar: Command Line Args** should now look something like this:

```
java -cp C:\Users\bbischoff\Desktop\karate-1.2.1.RC1\karate.jar
com.intuit.karate.Main
```

Let's now create a launch configuration that can be triggered by **Karate: Debug** in CodeLens and uses this setting.

Creating a launch configuration

To create a debug launch configuration, we can click on the **Karate: Debug** link in CodeLens above any test scenario. VS Code will then prompt us to select the debugger with which we want to run it.

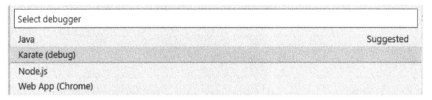

Figure 4.5 – Selecting a debugger

Here, we selected **Karate (debug)**. VS Code will now create and open the launch.json file that is located under your project's .vscode directory. This file contains one or more configurations for debugging.

You can see that in the file's `configuration` section, there is a name key with a value of Karate (debug): Standalone, indicating that Karate standalone will be used to run it.

Also, under `karateCli`, we see a reference to `${config:karateRunner.karateJar.commandLineArgs}`, which is exactly the setting we changed earlier.

```
 RUNNING-TESTS                          .vscode > {} launch.json > ...
   .vscode                          1   {
    {} launch.json                  2       // Use IntelliSense to learn about possible attributes.
    {} settings.json                3       // Hover to view descriptions of existing attributes.
    src\test\java              ⊛    4       // For more information, visit: https://go.microsoft.com/fwl
     examples                  ⊛    5       "version": "0.2.0",
      users                    ⊛    6       "configurations": [
       users.feature           U    7           {
       J UsersRunner.java       U    8               "type": "karate",
       J ExamplesTest.java      U    9               "name": "Karate (debug): Standalone",
    JS karate-config.js         U   10               "request": "launch",
    logback-test.xml            U   11               "feature": "${command:karateRunner.getDebugFile}",
   target                          12               "karateOptions": "",
    pom.xml                     U   13               "karateCli": "${config:karateRunner.karateJar.commar
                                    14           }
                                    15       ]
                                    16   }
```

Figure 4.6 – Karate debug launch configuration

Later in this chapter, we will add another configuration in the same file so we can choose between two options.

Now we should be able to use the **Karate: debug** CodeLens option successfully by clicking on it again. In the bottom-right corner of VS Code, there is an indicator that Karate's debug server is starting up and our debug session is connecting to it.

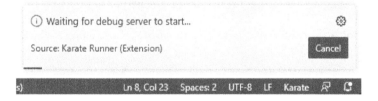

Figure 4.7 – Starting a debug server

It might not seem too useful right now because the test is running as before. The only difference, for now, can be seen in the logs.

Here, it is indicated both that it is running using Karate standalone and using the debug server:

```
Executing task: java -cp C:\Users\bbischoff\Desktop\karate-
1.2.1.RC1\karate.jar com.intuit.karate.Main -d
```

```
14:33:42.242 [main]  INFO  com.intuit.karate - Karate version:
1.2.1.RC1
14:33:43.000 [main]  INFO  com.intuit.karate.debug.DapServer -
debug server started on port: 52169
```

Let's put this setup to use!

Using the debug server

Especially during test development, it can be very interesting to find out what the current state of the test is at a specific step, which requests and responses were made in detail, and what would have happened if other values had been chosen. This is all possible with the debug server setup we have now.

Setting breakpoints

You can set breakpoints in the Karate scenario you want to debug. This means that execution will pause when the line with the breakpoint is reached.

To set a breakpoint, click on the left of the line where a break in execution should occur. Here, we set a breakpoint at line 10, Then status 200.

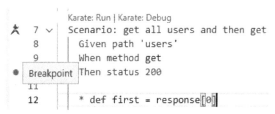

Figure 4.8 – Setting a breakpoint

The set breakpoint will also appear in the lower-left corner of VS Code along with the filename, path, and line number.

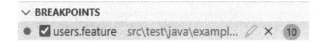

Figure 4.9 – Breakpoint list

It is also possible to set multiple breakpoints in case we want to see what is going on in more than one step.

Let's look at what happens when the test is debugged now.

Exploring the current state

Clicking **Karate: debug** with an active breakpoint stops execution when this line is reached. Also, it is highlighted to visualize this.

Figure 4.10 – Highlighted line with the reached breakpoint

On the left side of VS Code, this is reflected as well in the **CALL STACK** section.

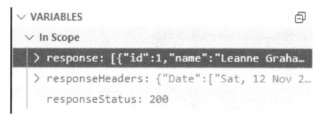

Figure 4.11 – VS Code call stack

The most important section is **VARIABLES**. Here, we can see all current variables that play a role in the current step.

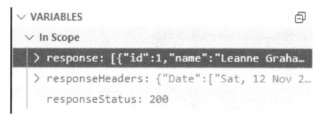

Figure 4.12 – VS Code breakpoint variables

In this case, we see `response JSON`, `responseHeaders`, and `responseStatus`, and we can even expand them to see the full contents. This makes debugging very straightforward.

Stepping back and forth

When a breakpoint is hit and execution stops, there is a new debug toolbar visible at the top of VS Code.

Figure 4.13 – The debug toolbar

Let's go through each of the buttons from left to right:

- **Continue**: This will continue the execution either to the next breakpoint or to the end if no other breakpoint is set.

- **Step Over**: This will pause at each subsequent line automatically, even if it doesn't have a breakpoint. However, if a line in the feature file executes an external method (as we will see in *Chapter 7, Customizing and Optimizing Karate Tests*), it will not go into this method.

- **Step Into**: This will pause at each subsequent line, also inside of external methods, automatically, even if they don't have breakpoints. If a line in the feature file executes an external method, it will also go through this method line by line.

- **Step Out**: When you are in an external method, stepping out will return right to the next scenario step and does not continue going through each line of the external method.

- **Step Back**: This goes back to the previous step and pauses.

- **Reverse**: In the Karate context, this does the same as **Step Back**.

- **Restart**: This hot-reloads your changes if you change steps while test execution is paused. The great thing about this is that we can also make changes *while the test is running*. You can step back to a previous test step and change parameters, URLs, variables, payloads, or assertions and continue forward again to see what changes.

- **Stop**: This stops the debug session and ends the debug server.

Also, it is possible to add or remove breakpoints during a debug session. This makes it a great tool for exploration and finding issues. Later in *Chapter 9, Karate UI for Browser Testing*, we will see that this also works in UI-based tests.

Using the debug console

There is a special **DEBUG CONSOLE** view in VS Code that allows us to further examine a running test.

Figure 4.14 – Debug console navigation link

We see the regular output that we also see in the terminal. However, additionally, when a test is paused through a breakpoint, we can type additional commands in the command line at the very bottom to explore the current state.

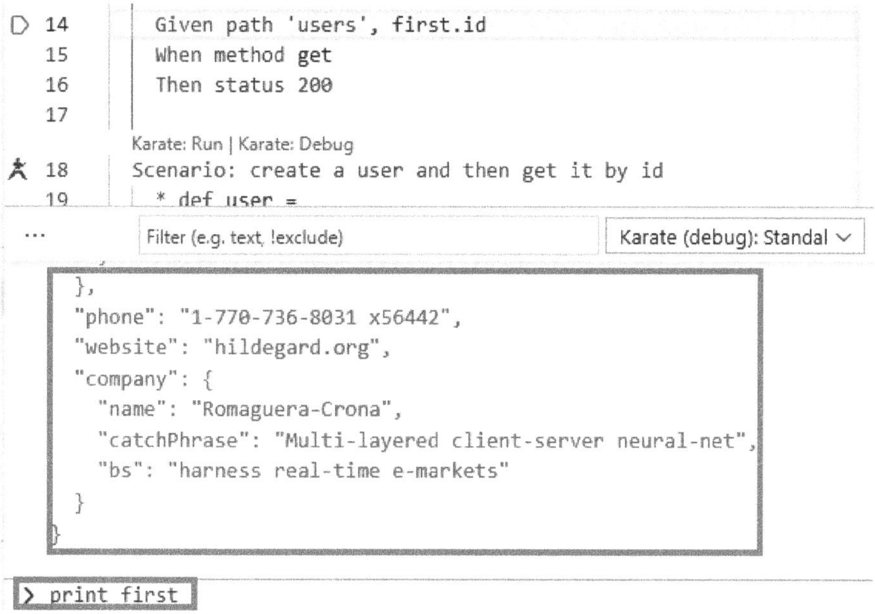

```
 ▷  14          Given path 'users', first.id
    15          When method get
    16          Then status 200
    17
                Karate: Run | Karate: Debug
 ⚐  18      Scenario: create a user and then get it by id
    19          * def user =
```

```
...     Filter (e.g. text, !exclude)                          Karate (debug): Standal ⌄
```

```
    },
    "phone": "1-770-736-8031 x56442",
    "website": "hildegard.org",
    "company": {
      "name": "Romaguera-Crona",
      "catchPhrase": "Multi-layered client-server neural-net",
      "bs": "harness real-time e-markets"
    }
}
```

```
> print first
```

Figure 4.15 – Debug console

Here, you can see that the breakpoint is hit in line 14 of the running test, which uses the `first` variable. If we want to see the current contents of it, we can either check **VARIABLES** as seen before or we can directly type the `print first` Karate command. This can then be executed directly by hitting the *Enter* key, and the result can be seen in the **Debug Console** window. This command line is very flexible and allows you to alter data, execute other Karate commands, and more.

In *Chapter 9*, *Karate UI for Browser Testing*, we will return to the debug command line again to see how we can use it while running tests in a web browser.

Now that we know how to use the debugger with Karate standalone, let's move on to see how to run Karate using Maven and runner classes.

Running tests with Maven

As we already have Maven set up to use for managing our project and dependencies, we will use it next to run the tests. This will be crucial for later when we want to set up Karate test runs within build pipelines. The most straightforward way to do this is to write one or more runner classes that Maven can execute. In this section, we will look at the runners that are already included in our generated Maven archetype example project.

Understanding Karate runners

In our project, we can see two different files that are examples of how to run tests:

1. `ExampleTests.java` in the `examples` directory
2. `UsersRunner.java` right next to `users.feature` inside of `examples/users`

Figure 4.16 – Included runners in Karate's example project

Runners are needed when tests should be run from Maven, more specifically **Maven Surefire**. This is the default Maven plugin for executing Java unit tests.

Looking at the `pom.xml` file in our project, there is a specific plugin dependency defined for this:

```
<plugin>
    <groupId>org.apache.maven.plugins</groupId>
    <artifactId>maven-surefire-plugin</artifactId>
    <version>${maven.surefire.version}</version>
    <configuration>
        <argLine>-Dfile.encoding=UTF-8</argLine>
    </configuration>
</plugin>
```

The version number is specified further up in the same files under the **properties** section:

```
<maven.surefire.version>2.22.2</maven.surefire.version>
```

When we execute the `mvn test` command from the VS Code terminal or the command prompt of the operating system, this will automatically invoke Surefire, which in turn runs **JUnit** tests.

> **A word about JUnit**
>
> JUnit is the standard Java unit testing framework, which is also part of our Karate dependency (we are specifying `karate-junit5` in the Maven `pom.xml` file). Surefire runs JUnit tests out of the box and thus also Karate tests. This makes it possible to easily run these from the IDE as well as the command line as we will see next.

Let's look at both runner classes in more detail, find out what they contain, and understand how they are used.

Understanding UsersRunner.java

Let's look at UsersRunner.java first. It looks like this:

```
package examples.users;
import com.intuit.karate.junit5.Karate;

class UsersRunner {
    @Karate.Test
    Karate testUsers() {
        return Karate.run("users").relativeTo(getClass());
    }
}
```

This class uses the special @Karate.Test Karate **JUnit 5** annotation that marks a test method. This method (which is called testUsers() here but could have any name) contains the code to run the users feature. relativeTo(getClass()) indicates that Karate should look for the users feature file within the package of the runner class.

We can execute the runner method right from VS Code as well to verify this. It will execute the users.feature file correctly.

```
 7        @Karate.Test
 8        Karate testUsers() {
 9            return Karate.run(...paths: "users").relativeTo(getClass());
10        }
```

Figure 4.17 – Executing the runner method

A left-click on the green arrow executes the test and a right-click reveals more options, such as running or debugging the test. Additionally, you can execute your runner classes and methods through the **Testing** tab of VS Code.

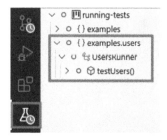

Figure 4.18 – VS Code Testing tab

We will see later in this chapter that we can employ a runner with different methods to select exactly which tests we want to run.

Executing UsersRunner through Maven

To use this runner from the command line, we can use the `test` Maven command. You can use the `test` parameter to specify which runner you want to use. In the case of `UsersRunner`, the command looks like this:

```
mvn test -Dtest=UsersRunner
```

Parameters in Maven are set using the `-D` option. This command should work both in VS Code's terminal window as well as the operating system command prompt.

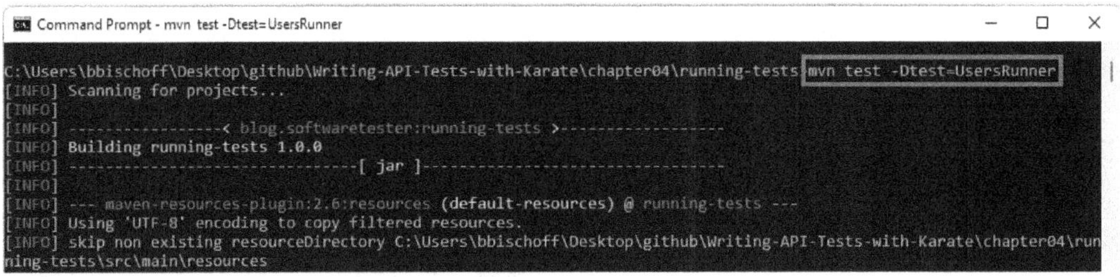

Figure 4.19 – Executing a runner class using Maven on the command line

Note that it is only necessary to state the name of the runner class without the `.java` file extension or any specific path. We will see later in this chapter how we can further use this mechanism.

It might happen that you get a firewall security alert when running the test like this for the first time. If this occurs, it should be enough to choose **Private networks, such as my home or work network** to fix this.

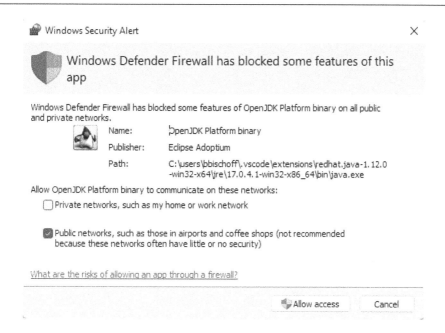

Figure 4.20 – Windows security alert

Let's look at the other runner in our example project next and what is different about it.

Understanding ExampleTest.java

In the `examples` package of the project, we can find the `ExampleTest.java` class. Let's go through the different parts of it and see what they are there for:

```
package examples;

import com.intuit.karate.Results;
import com.intuit.karate.Runner;
import static org.junit.jupiter.api.Assertions.*;
import org.junit.jupiter.api.Test;

class ExamplesTest {
    @Test
    void testParallel() {
        Results results = Runner.path("classpath:examples")
                //.outputCucumberJson(true)
```

```
                    .parallel(5);
        assertEquals(0, results.getFailCount(), results.
 getErrorMessages());
    }
}
```

This is a little more complex than the previous runner class, but it is also more flexible:

- The first difference is that it does not use the @Karate.Test annotation but rather the native JUnit @Test annotation. This means that this is using pure JUnit functionality to execute the testParallel() annotated test method.

- The next noteworthy point is the definition of Runner.path("classpath:examples"), which means that this runner would run everything included in the examples package. This works because of classpath, which contains all existing test classes and resources and enables JUnit to find these automatically.

- The commented-out part (.outputCucumberJson(true)) will play a role later in *Chapter 5, Reporting and Logging*, and is not important right now.

- The .parallel(5) part specifies the number of threads that Karate should allocate to run tests in parallel, in this case, 5. UsersRunner did not have this capability before and executed all tests sequentially (meaning one after the other). The parallel command also *runs* the tests and returns a list of results that are then saved in the results variable.

   It is important to note that Karate does not necessarily execute tests in the same order every time that they are executed in parallel, as they are started as soon as there is a free *slot* in one of the threads.

- The last crucial mechanism is the assertEquals(0, results.getFailCount(), results.getErrorMessages()); assertion. This part determines the number of failed tests and passes the whole test suite if 0 failures occurred. If there are failures, it prints out a list of all error messages that were collected in the results list.

   The reason behind this is that due to the parallel execution of test cases, JUnit needs to wait for all *individual* test results before it can be sure that all tests passed.

Let's see how we can run this ExampleTest runner using Maven.

Executing the ExampleTest runner through Maven

Right now, we can use the ExampleTest runner class from Maven as before from the command line:

```
mvn test -Dtest=ExamplesTest
```

We can see in the logs that two scenarios were executed with a maximum number of 5 threads (in this case, only 2 threads were used, of course):

```
===========================================================
elapsed:    5.46 | threads:      5 | thread time: 2.35
features:      1 | skipped:      0 | efficiency: 0.09
scenarios:     2 | passed:       2 | failed: 0
===========================================================
```

The interesting thing is that even if we run mvn test without any test parameter, this still works. This is due to Surefire's convention: it will automatically run all classes that end with *Test* if no explicit class is specified. This can be very convenient but also really confusing if you don't know about this.

Let's look at yet another way to run the ExamplesTest runner inside our IDE.

Executing the ExampleTest runner in the IDE

As before, we can easily run the new runner from the IDE by clicking the green run button beside it.

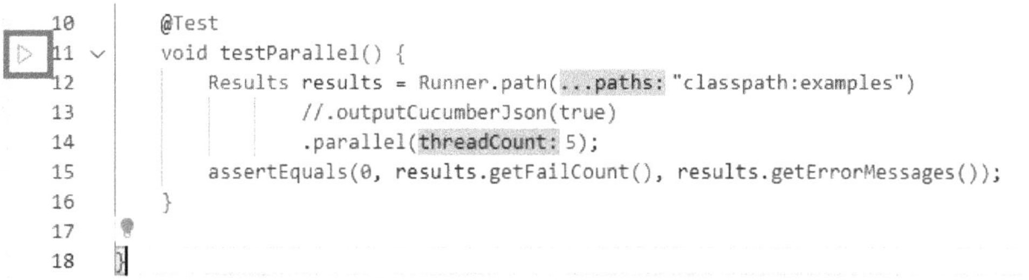

```
10      @Test
11  ∨   void testParallel() {
12          Results results = Runner.path(...paths: "classpath:examples")
13                  //.outputCucumberJson(true)
14                  .parallel(threadCount: 5);
15          assertEquals(0, results.getFailCount(), results.getErrorMessages());
16      }
17
18  }
```

Figure 4.21 – Executing the parallel runner in VS Code

Let's see next how we can use the Maven runner and choose which scenario to run.

Executing a Maven Karate runner

If the **Karate Cli: Override Karate Runner** setting is still active, we need to deactivate it.

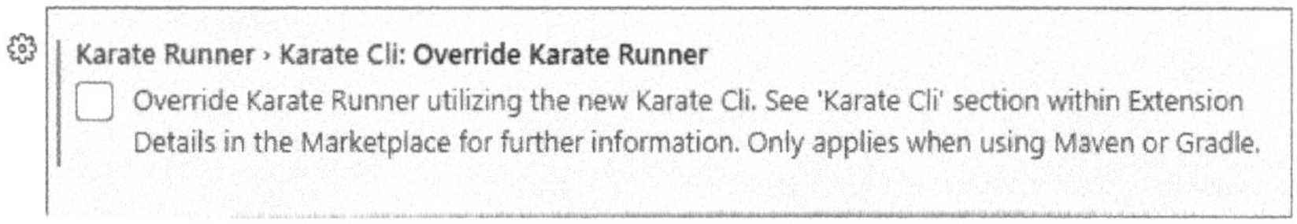

Karate Runner › Karate Cli: Override Karate Runner
☐ Override Karate Runner utilizing the new Karate Cli. See 'Karate Cli' section within Extension Details in the Marketplace for further information. Only applies when using Maven or Gradle.

Figure 4.22 – Deactivating the Karate runner CLI override

Now if we try to run a scenario via **Karate: Run** in CodeLens, we should get a prompt for a Karate runner.

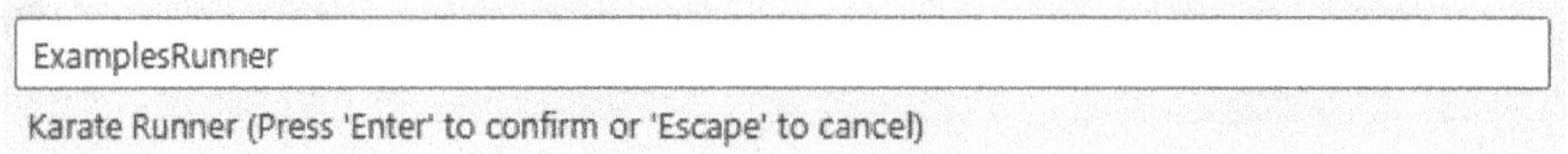

```
ExamplesRunner
```
Karate Runner (Press 'Enter' to confirm or 'Escape' to cancel)

Figure 4.23 – Specifying a Karate runner

Here, we can enter ExamplesTest because this name is passed to the new Maven run configuration. Confirming by pressing the *return* key runs the chosen test using the specified runner.

We can see this in the logs. Let's dissect this command into its parts:

1. The first part of the command is the standard mvn clean test command with a reference to our test project's pom.xml file:

    ```
    Executing task: mvn clean test -f""c:\Users\bbischoff\
    Desktop\github\Writing-API-Tests-with-Karate\chapter04\
    running-tests\pom.xm""
    ```

2. Next, we see the ExamplesTest runner class as we specified before on the command line:

    ```
    -Dtest=ExamplesTest
    Finally, karate.options are specified which point to the
    feature file and line number of the test scenario we want
    to run (users.feature, line 7):""-Dkarate.options=c:\
    Users\bbischoff\Desktop\github\Writing-API-Tests-with-
    Karate\chapter04\running-tests\src\test\java\examples\
    users\users.feature:""
    ```

After running once with this runner, this is then automatically configured as the standard runner for this project. VS Code will still ask us every time which runner we want to use, but ExamplesTest will be prefilled in the **Karate Runner** dialog. If desired, you can also deactivate the **Karate Runner: Prompt To Specify** option so it will run using the standard runner class right away.

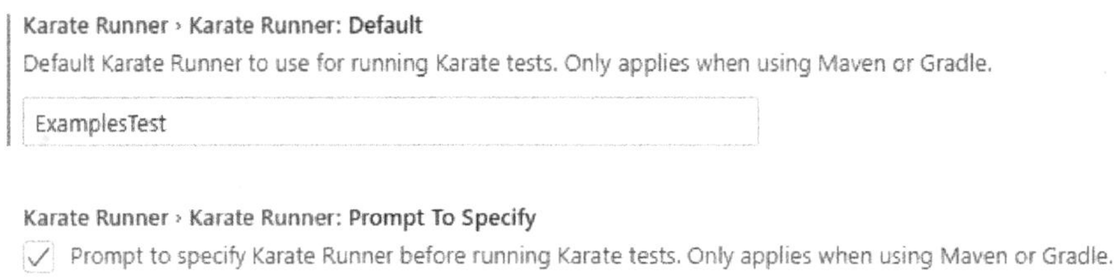

Karate Runner › Karate Runner: **Default**
Default Karate Runner to use for running Karate tests. Only applies when using Maven or Gradle.

```
ExamplesTest
```

Karate Runner › Karate Runner: **Prompt To Specify**
✓ Prompt to specify Karate Runner before running Karate tests. Only applies when using Maven or Gradle.

Figure 4.24 – Default Karate runner options

Now that we know some different ways to run tests, let's check how to run them against different environments.

Running tests against different environments

Many companies have different environments that are used during the software delivery life cycle.

It starts with the local machines on which the applications are developed. This is often followed by staging, which in the best case is an environment that is as close as possible to the production environment. Between staging and production environments, there may be other steps, for example, for user acceptance testing.

The problem here is that each environment has a different base URL, sometimes even a different endpoint-, up to a completely different access method and configuration.

Karate offers a very straightforward mechanism to cope with this. To demonstrate this, we created a new demo project called `environments`. The code for this example can be found at `https://github.com/PacktPublishing/Writing-API-Tests-with-Karate/tree/main/chapter04/environments`.

Figure 4.25 – Environment demo project setup

The next sections explain the mechanism for environment-specific properties.

Using a custom property in a feature file

This feature file demonstrates using a custom `baseUrl` property that is not set within the scenario itself. Instead, it will come from `karate-config.js` based on the passed environment string. For this demonstration, it just prints the current value of `baseUrl` to illustrate the concept:

```
Feature: Feature 1
  Scenario: Choosing environments
    * print "Base URL", baseUrl
```

Note that there is no complicated setup necessary to retrieve the custom baseUrl property!

Using Karate's environment property

Karate has a special property called karate.env that is used to pick an environment. This can either be set within a runner (so that we can have different runner methods for different environments), or it can be passed as a system property to the test run.

The runner methods could look like this (this is taken from the Runner.java class inside the environments example project):

```
@Karate.Test
Karate dev() {
    // Runs the scenario with the dev environment
    return Karate.run("env-demo").karateEnv("dev")
        .relativeTo(getClass());
}


@Karate.Test
Karate prod() {
    // Runs the scenario with the prod environment
    return Karate.run("env-demo").karateEnv("prod")
        .relativeTo(getClass());
}
```

You can also do the exact same thing in a pure JUnit runner:

```
@Test
void testProd() {
    Results results =
        Runner.path("classpath:tests")
            .karateEnv("prod").parallel(1);
    assertEquals(0, results.getFailCount()
        results.getErrorMessages());
}
```

To run either of those methods via Maven, we can specify the runner class and the name of the method after a # character:

```
mvn test -Dtest=Runner#dev
```

Or we can do as follows:

```
mvn test -Dtest=Runner#prod
```

If the `.karateEnv()` option is not used within the runner method, we can also pass it through the `karate.env` property, as in the following example:

```
mvn test -Dtest=Runner#runnerMethod -D"karate.env"=prod
```

Note that `karate.env` must be surrounded by quotes; otherwise, Maven might have difficulty accepting a property containing a dot.

Setting up karate-config.js

The `karate-config.js` optional configuration file is in the `src/test/java` directory. It can hold centralized variables and functionality that should be available to all test scenarios. Whenever you need a centralized source of truth that holds variables that are needed by all tests, this is a very convenient way to solve this.

A `karate-config.js` file typically contains one JavaScript function that is automatically called at the beginning of each scenario. It should return an object called `config` that contains key-value pairs of properties. We will look at this more deeply in *Chapter 6, More Advanced Karate Features*.

For now, our goal is to set a base URL for our scenarios depending on the test environment. This will simulate the use case mentioned before.

Our custom `karate-config.js` looks like this:

```
function fn() {
  var env = karate.env;
  karate.log('karate.env system property is:', env);

  if (!env) env = 'dev';
  var config = { env: env }

  if (env == 'dev') {
    config.baseUrl = "devBaseUrl"
  } else if (env == 'prod') {
    config.baseUrl = "prodBaseUrl"
  }
  return config;
}
```

This is what happens here:

1. We create an env variable and set it to the special `karate.env` property.

2. If `karate.env` was not passed, we set it to *dev* as a fallback.

3. We create a config object and add an env property with the value of the env variable in case we need to access it later.

4. Based on the value of env, we can then add the correct `baseUrl` to the `config` object.

5. In the end, we return the `config` object. At this point, all properties included in this object are automatically made available to all test scenarios!

Running `mvn test -D"karate.env"`=prod will now give us this output:

```
22:29:27.236 [main] INFO  com.intuit.karate.Runner - using
system property 'karate.env': prod

…

22:29:29.497 [main] INFO  com.intuit.karate - karate.env system
property is: prod
22:29:29.563 [main] INFO  com.intuit.karate - [print] Base URL
prodBaseUrl
```

If we run it with `-D"karate.env"`=dev, we correctly see the following:

```
22:29:27.236 [main] INFO  com.intuit.karate.Runner - using
system property 'karate.env': dev

…

22:29:29.497 [main] INFO  com.intuit.karate - karate.env system
property is: dev
22:29:29.563 [main] INFO  com.intuit.karate - [print] Base URL
devBaseUrl
```

Now that we have seen how to run tests against different environments, let's now learn how we can run selected tests to streamline our test suite and tailor it to the needs of the project.

Running specific tests

During the software development life cycle, there is rarely a case when you want to run all existing tests in your Karate project; for example, when a full regression test is necessary on your last environment prior to production deployment. Most of the time, it is sufficient to only run a smaller set of tests that fit the current use case.

Examples of this could be as follows:

- During feature development, only the tests associated with the feature under development are run

- During test development, only the current test under development is run to make sure it tests what it should

- After deployment, a few smoke-test scenarios are run to verify that it was successful

- A small set of tests covering the most important flows are run against the live instances of APIs for monitoring purposes

This is of course completely dependent on your or your team's requirements.

To make it easier to show this, we created a new demo project called `choosing-tests`. The code for this example can be found at `https://github.com/PacktPublishing/Writing-API-Tests-with-Karate/tree/main/chapter04/choosing-tests`.

This project contains two scenarios: `tests/Test1.feature` and `tests/Test2.feature`. Both are set up in the same way. They each contain two scenarios that just print out their name so we can better see what happens:

```
Feature: Feature 1
  Scenario: Feature 1 Scenario 1
    * print "Feature 1 Scenario 1"


  Scenario: Feature 1 Scenario 2
    * print "Feature 1 Scenario 2"
```

Let's now look at some ways to select specific tests or test suites to run.

Running specific runner class methods

We already saw that a runner class can be used to execute specific tests and even specific environments. As a recap, you can pick the runner class and method when calling `mvn test`:

```
mvn test -Dtest=RunnerClass#runnerMethod
```

This enables us to create separate methods that are named appropriately and execute a subset of tests for specific purposes.

If you consider these runner classes, it makes it clearer:

```
public class Runner {
    @Karate.Test
```

```
Karate test1() {
    // Runs all scenarios in test1.feature
    return Karate.run("test1").relativeTo(getClass());
}


@Karate.Test
Karate test2() {
    // Runs all scenarios in test2.feature
    return Karate.run("test2").relativeTo(getClass());
}


@Karate.Test
Karate test1and2() {
    // Runs all scenarios in test1 and test2.feature
    return Karate.run("test1", "test2")
        .relativeTo(getClass());
}
}
```

Running `mvn clean test -Dtest=Runner#test2`, for example, gives us this output:

```
[INFO] Running tests.Runner
21:27:19.247 [main] INFO  com.intuit.karate - [print] Feature 2
Scenario 1
21:27:19.307 [main] INFO  com.intuit.karate - [print] Feature 2
Scenario 2
```

We can see this even better in the **DEBUG CONSOLE** view where the run feature is clearly stated.

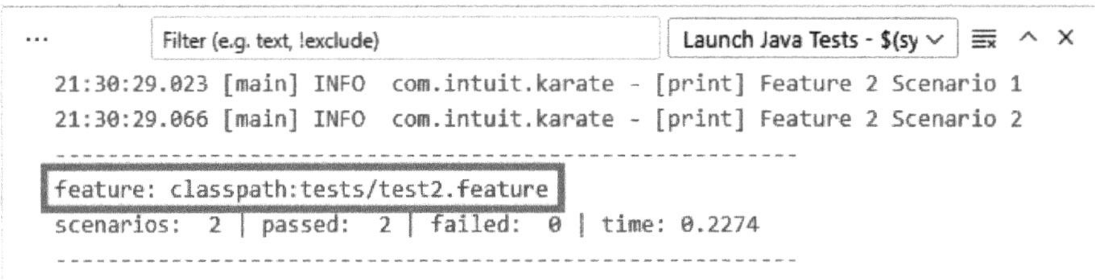

Figure 4.26 – VS Code debug console view

These debug console logs only appear when starting the test from inside the IDE and not via the Maven command in the terminal!

Another nice feature of VS Code in this regard is that we can see in the **Testing** view what features belonged to a specific run method after we execute it once from the IDE. This can then also be used to rerun specific runners and tests.

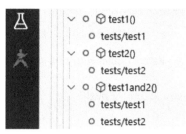

Figure 4.27 – Runners and features in VS Code's Testing view

As you can see, we can also specify a list of features, as in the following example:

```
Karate.run("test1", "test2")
```

Luckily, the exact same ways work also for the parallel runner, but we just need to specify the test paths a little differently:

```
public class ParallelTest {
    @Test
    void testAll() {
        Results results =
            Runner.path("classpath:tests").parallel(4);
        assertEquals(0, results.getFailCount(),
            results.getErrorMessages());
    }

    @Test
    void testFeature1() {
        Results results =
        Runner.path("classpath:tests/Test1.feature")
            .parallel(4);
        assertEquals(0, results.getFailCount(),
        results.getErrorMessages());
    }
}
```

It is good to know that in this case, when we have such a runner with multiple methods, we can also specify a specific one when we use **Karate: Run** from CodeLens.

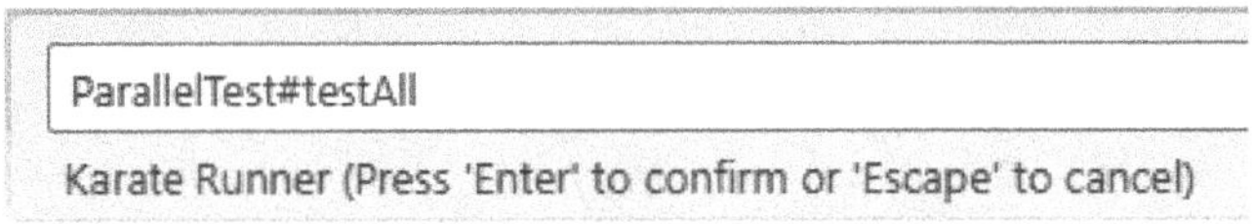

ParallelTest#testAll

Karate Runner (Press 'Enter' to confirm or 'Escape' to cancel)

Figure 4.28 – Karate runner with a specific method

Let's quickly look at how to run chosen scenarios next.

Running specific scenarios

We will see how to select scenarios:

- Inside a runner
- From the command line

Selecting scenarios inside a runner

Running a specific scenario inside a feature file just requires you to append the line number of the scenario after a colon (:). If you want to run the second scenario that starts on line 6 in Test1. feature, you can call it like this:

```
@Test
void testFeature1Scenario2() {
    Results =
    Runner.path("classpath:tests/Test1.feature:6")
        .parallel(4);
    assertEquals(0, results.getFailCount(),
    results.getErrorMessages());
}
```

Selecting scenarios from the command line

If you want to run specific feature files from the command line, there is a way that we already used implicitly when running tests through VS Code. This is using the special karate.options property, which can be used to set which scenarios you want to run.

If you use the parallel runner from before, for example, you can do it using a command like this:

```
mvn clean test -Dtest=ParallelTest#testAll
    "-Dkarate.options=src\test\java\tests\Test1.feature:6"
```

This must be written in *one line*; otherwise `karate.options`, will be ignored!

It allows us to select features or scenarios without altering the runner class and methods. This comes in handy when using Karate runs from the command line within CI/CD pipelines. It is noteworthy here that the path to the features to run should be relative to the directory of the `pom.xml` file or a full absolute path. You can also use the `classpath:` prefix to select a feature within `classpath` and omit the concrete path.

Let's look at how we can further filter tests so we can create more meaningful test suites.

Filtering tests by tags

Tags can be placed on scenarios or whole feature files to pick which ones to run, which ones not to run, or combinations of both. Tags can even be used to group scenarios regardless of which feature they reside in.

To illustrate this, we added some tags to the existing `Test1.feature` files like so:

```
Karate: Run | Karate: Debug
Feature: Feature 1
    Karate: Run | Karate: Debug
    @smoke @important
    Scenario: Feature 1 Scenario 1
      * print "Feature 1 Scenario 1"


    @ignore
    Scenario: Feature 1 Scenario 2
      * print "Feature 1 Scenario 2"
```

Figure 4.29 – Feature 1 with tags

The first scenario has a `@smoke` and an `@important` tag whereas the second one has an `@ignore` tag. This `@ignore` tag is special as these tagged tests are not run. The VS Code plugin even deactivates the CodeLens options for this scenario entirely to illustrate this.

```
Karate: Run | Karate: Debug
Feature: Feature 2
    Karate: Run | Karate: Debug
    Scenario: Feature 2 Scenario 1
      * print "Feature 2 Scenario 1"


    Karate: Run | Karate: Debug
    @smoke
    Scenario: Feature 2 Scenario 2
      * print "Feature 2 Scenario 2"
```

Figure 4.30 – Feature 2 with tags

The `Test2.feature` file's first scenario has no tags whereas the second one also has the `@smoke` tag.

Running this command now from the command line picks all scenarios that have the `@smoke` tag:

```
mvn clean test -Dtest=ParallelTest#testAll
    "-Dkarate.options=classpath: --tags=@smoke"
```

As before, this command must be one line!

Here we use the special `--tags` flag of `karate.options`. Also, we use an empty `classpath:` definition to choose all feature files inside `classpath` automatically.

We can see that this only runs the two scenarios that are tagged with `@smoke`:

```
20:04:27.502 [main] INFO  com.intuit.karate.Runner - using
system property 'karate.options': classpath: --tags=@smoke
20:04:28.536 [main] DEBUG com.intuit.karate.Suite - waiting for
2 features to complete
20:04:29.953 [pool-1-thread-1] INFO  com.intuit.karate -
[print] Feature 1 Scenario 1
20:04:29.954 [pool-1-thread-2] INFO  com.intuit.karate -
[print] Feature 2 Scenario 2
```

Karate also prints out `karate.options` that were passed, which makes it easy to see what is executed and whether we made any mistakes.

You can also specify that you don't want to run specific tags by prefixing them with a tilde (~):

```
mvn clean test -Dtest=ParallelTest#testAll
    "-Dkarate.options=classpath: --tags=~@important"
```

As a final remark, tags can also be specified within the runner methods if this is desired. It can be beneficial, for example, to have a specific method just to run smoke tests:

```
@Karate.Test
Karate testSmokeTag() {
    // Runs all scenarios tagged with @smoke
    return Karate.run()
        .relativeTo(getClass()).tags("smoke");
}
```

This concludes our discussion on tags for now. In *Chapter 6, More Advanced Karate Features*, we will see more uses of tags that go beyond simple filtering, such as environment-specific tagging and tags with values.

Summary

In this chapter, we looked at some different ways to run Karate tests, both from inside and outside the IDE. Also, we learned how to debug tests to find bugs early on. We took closer looks at running tests via Maven, how to parallelize tests to save time, and checked out various ways of running only a specific subset of tests using line numbers and tags.

In *Chapter 5*, *Reporting and Logging*, we will look at Karate's reporting, how to configure it, and explore how it can help the different stakeholders to make the most of the test results.

Later on in *Chapter 8*, *Karate in Docker and CI/CD Pipelines*, we will revisit the topic of running Karate tests and look at how we can run Karate in these environments.

5
Reporting and Logging

In the previous chapter, we learned about several ways to run tests, but we didn't see the results of those tests in a representative and comprehensible way. Looking at the log outputs is a good way to analyze what's going on during a test, but it's not something we want to give to the non-technical people on the team to read.

We'll start by looking at some ways to improve the readability of the logs, as this will make it easier for you, as a test developer, to find problems. In addition, Karate has great reporting features that can help us, as well as anyone interested in the test results, to better understand them and know exactly what is being tested.

In this chapter, we will cover these main topics:

- Configuring log options
- Using Karate's built-in reports
- Configuring third-party report libraries
- Generating a Surefire report
- Using JUnit report

Technical requirements

The code examples for this chapter can be found in this book's GitHub repository: `https://github.com/PacktPublishing/Writing-API-Tests-with-Karate/tree/main/chapter05`.

You will require the system and IDE setup we completed in *Chapter 2, Setting Up Your Karate Project*, for this chapter.

First, let's learn how we can configure Karate's logging options.

Configuring log options

When working on new tests or debugging existing ones, the logs are usually a good indicator of what is going on while a test is being run and what happens if it fails. While we see them on the command line when running a test there, they are also usually stored within the Maven project's `target` directory – the standard directory that is created when running the test project via Maven:

Figure 5.1 – karate.log within the target directory

Now, let's look at log levels and what they mean!

Understanding log levels

Log levels determine which kinds of logs you want to include in the output. Karate uses the logging library **Logback**, which supports multiple log levels: TRACE, DEBUG, INFO, WARN, and ERROR. These get less verbose from left to right and determine the scope and purpose of each log entry. The most common one is INFO, which indicates a log output that prints out the basic actions and values that are needed to understand the application flow.

A certain log level always *excludes all lesser* but *includes all higher* log levels. For example, setting the log level to TRACE would include *all* log levels, whereas setting it to WARN would only show WARN and ERROR. These are generally shown in front of each line of logs as well:

```
[INFO] Results:
[INFO]
[ERROR] Failures:
[ERROR]    Run.testParallel:14 status code was: 404, expected: 200,
```

Figure 5.2 – Log levels

Now, let's look at the default log configuration.

Examining the Logback configuration

You might have seen that, in the Maven archetype-generated projects, there is a file called `logback-test.xml` that typically resides inside the `src/test/java` directory. This file looks like this:

```xml
<?xml version="1.0" encoding="UTF-8"?>
<configuration>
    <appender name="STDOUT" class="ch.qos.logback.core.
ConsoleAppender">
        <encoder>
            <pattern>%d{HH:mm:ss.SSS} [%thread] %-5level
%logger{36} - %msg%n</pattern>
        </encoder>
    </appender>
    <appender name="FILE" class="ch.qos.logback.core.
FileAppender">
        <file>target/karate.log</file>
        <encoder>
            <pattern>%d{HH:mm:ss.SSS} [%thread] %-5level
%logger{36} - %msg%n</pattern>
        </encoder>
    </appender>
    <logger name="com.intuit" level="DEBUG"/>
    <root level="info">
        <appender-ref ref="STDOUT" />
        <appender-ref ref="FILE" />
    </root>
</configuration>
```

It defines two appender elements, which are responsible for writing logs to specific destinations:

1. The STDOUT appender takes care of the console output.
2. The FILE appender logs to the `target/karate.log` file.

Both appenders include a `<pattern>` element that defines the formatting for log lines. In this configuration, both the console and file logger use the same pattern. If you go to `https://logback.qos.ch/manual/layouts.html#ClassicPatternLayout`, you can read more about what options there are and how you can customize them.

Next in line is the `<logger name="com.intuit" level="DEBUG"/>` element, which essentially states that everything in the `com.intuit` package (the root package of Karate itself) should be logged with the `DEBUG` level. Otherwise, everything else should be logged with `INFO`, as configured by `<root level="info">`. This so-called *root logger* includes the references to the two appenders, meaning that both console and file logging are enabled at the same time.

Changing the log level

Currently, the logs look like this when running the tests:

```
21:11:56.304 [pool-1-thread-1] DEBUG com.intuit.karate - response time in mill
iseconds: 316
1 < 404
1 < Date: Tue, 29 Nov 2022 21:12:33 GMT
1 < Content-Type: application/json; charset=utf-8
1 < Content-Length: 2
1 < Connection: keep-alive
1 < X-Powered-By: Express
1 < X-Ratelimit-Limit: 1000
```

Figure 5.3 – Debug logs

This means Karate logs requests and responses using the `DEBUG` level. This can be helpful when debugging tests but overwhelming when going through production logs.

To change this, we just need to change the log level to a higher one in the Karate internal logger, depending on what we want to see:

```
<logger name="com.intuit" level="INFO"/>
```

Here, I set the level to `INFO` because I am interested in the `INFO`, `WARNING`, and `ERROR` logs. Running the tests again, we should not see the request and response logs from before:

```
22:18:57.820 [main] INFO com.intuit.karate.Suite - backed up existing 'target
\karate-reports' dir to: target\karate-reports_1669756737815
22:19:01.109 [pool-1-thread-2] ERROR com.intuit.karate - classpath:reporting/d
emo/Reporting.feature:23
Then status 200
status code was: 404, expected: 200, response time in milliseconds: 277, url:
https://jsonplaceholder.typicode.com/this_is_wrong, response:
{}
classpath:reporting/demo/Reporting.feature:23
```

Figure 5.4 – Suppressed DEBUG logs

Next let's take a look at how to supress print steps.

Suppressing print

It can be beneficial to enable print statements only, depending on the environment you are testing on or while you are developing tests. By default, `*  print` steps are shown in the logs but they can be suppressed in two ways:

- Inside a scenario, you can use Karate's `configure` method to disable it, like this:

  ```
  * configure printEnabled = false
  ```

  This will not show any `print` statements from that point on.

- In the `karate-config.js` file, you can add the following to the configuration method, which sets this for all tests:

  ```
  function fn() {
     karate.configure("printEnabled", false);
     var config = {
     }
     return config;
  }
  ```

Now that we have seen the different log configurations, let's move on to Karate's internal reports, what they show, and how they can be configured.

Using Karate's built-in reports

Karate has a very good reporting functionality that shows what was run, which steps were executed, and what was done within the steps. Also, it gives you information about the used tags and how tests are distributed across different threads when run in parallel.

When running a test suite, we typically get the following output at the end of the test run. This points to the report that Karate automatically generates:

```
HTML report: (paste into browser to view) | Karate version:
1.2.0


file:///C:/Users/bbischoff/Desktop/github/Writing-API-Tests-
with-Karate/chapter05/karate-reports/target/karate-reports/
karate-summary.html
```

As you can see, Karate generates the report in Maven's `target` directory in a new `karate-reports` folder.

To view this report, you can either copy and paste this into a web browser or switch to the **Karate** tab of the VS Code Karate plugin:

Figure 5.5 – Karate report links in the VS Code Karate tab

Under **REPORTS**, you will find separate links that directly open the different pages of the report in your default web browser.

The different report pages

Karate's report has four different pages. Let's quickly look at these here.

Summary

This page shows the feature overview. Here, you can see the names of features, how many scenarios were run in each, and how many of these passed and failed:

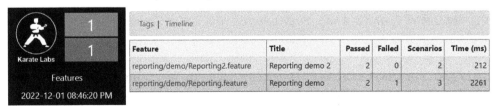

Feature	Title	Passed	Failed	Scenarios	Time (ms)
reporting/demo/Reporting2.feature	Reporting demo 2	2	0	2	212
reporting/demo/Reporting.feature	Reporting demo	2	1	3	2261

Figure 5.6 – Karate feature overview

You can click on a feature to see all its included scenarios in the left sidebar of the report:

Figure 5.7 – Scenarios within a specified feature

The right-hand side shows all the steps of the scenarios.

Scenario

Upon clicking on a scenario, you are taken directly to its steps. You can see all the steps that were executed and, in case of a failure, the exception regarding what went wrong:

failing	ms: 106
Scenario: [3:19] **wrong path**	
20 Given url 'https://jsonplaceholder.typicode.com'	0
21 And path 'this_is_wrong'	4
22 When method get	101
23 Then status 200	0

```
20:46:17.924 classpath:reporting/demo/Reporting.feature:23
Then status 200
status code was: 404, expected: 200, response time in milliseconds: 97, url:
https://jsonplaceholder.typicode.com/this_is_wrong, response:
{}
classpath:reporting/demo/Reporting.feature:23
```

Figure 5.8 – Exceptions in Karate's built-in report

On the first line, you can see the scenario's name, the column and line number of where it is located inside the feature (in this case, [3:19] stands for *3rd line, 19th column*), and the tags (here, this is failing).

It is also useful that the requests and responses can be seen when clicking on one of the steps that performs these operations. This works when the log level of com.intuit is set to DEBUG:

```
22 When method get

20:46:17.825 request:
1 > GET https://jsonplaceholder.typicode.com/this_is_wrong
1 > Host: jsonplaceholder.typicode.com
1 > Connection: Keep-Alive
1 > User-Agent: Apache-HttpClient/4.5.13 (Java/17.0.4.1)
1 > Accept-Encoding: gzip,deflate
```

Figure 5.9 – Request details in Karate's built-in report

This is typically the view that is most important when analyzing the status of a specific test suite run.

Tags

This page shows the distribution of your used tags among the scenarios. This is presented as a table with all tags. Each feature file that uses one of these tags has an **X** in the respective column:

Feature	failing	passing
reporting/demo/Reporting2.feature		X
reporting/demo/Reporting.feature	X	X

Figure 5.10 – Karate report tag distribution

Timeline

This view is interesting if you want to know the inner workings of the parallel runs. It shows all the different threads that were used in the test run, along with the scenarios that were run on each:

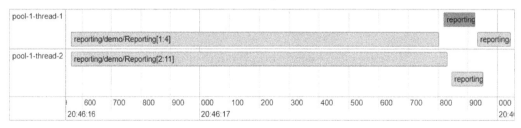

Figure 5.11 – Karate report parallel test run visualization

In this example, there are two threads (`pool-1-thread-1` and `pool-1-thread-2`). The first one ran three scenarios, whereas the second ran two. You can even see the feature's filename and the line numbers of each scenario and their runtime.

> **Ramping up the framework**
>
> You will notice that the first test scenario that's executed in each thread takes much longer than the ones after. This is due to the framework starting up when running them. After these two scenarios finish, the ones after run much more quickly.

Let's look at some options for configuring the internal reports further.

Preserving old reports

You might have noticed that there can be multiple reports under `target`:

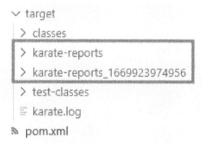

Figure 5.12 – Karate report backup

This is Karate's standard behavior. Each new report is saved under **karate-reports**, whereas already existing ones are moved to a new folder starting with `karate_reports` and ending with a timestamp. These can be helpful to check progress while developing new tests or seeing the history of test runs. However, those reports only exist if the target directory is not deleted (for example, by calling `mvn clean`).

If you want to change this behavior, you can configure it by setting the `backupReportDir` property to `false` inside the runner:

```
Results = Runner.path("classpath:reporting")
    .backupReportDir(false)
    .parallel(2);
```

Deciding what to report

Reporting all the information we have is nice for debugging while authoring new tests. For production, it is more helpful to only show relevant information to make test analysis easier.

Suppressing request and response display

Karate displays requests and responses in its reports by default. This can be turned off if this is not desired by using the showLog option. Like all configuration options, this can be used in a scenario step like this:

```
* configure report = {showLog: false}
```

Alternatively, you can add this to the karate-config.js file:

```
karate.configure("report", {showLog: false});
```

Suppressing helper steps

To further clean up reports, it is also possible to not include all steps starting with *. This is especially useful when these steps just create temporary variables and calculations that are not relevant to the viewer of test results. For this, you can add the showAllSteps: false JSON key to the report configuration:

```
* configure report = {showLog: true, showAllSteps: false}
```

This configuration would show requests and responses but no * steps.

Now, let's move away from Karate's internal reports and see how we can integrate other reporting libraries.

Configuring third-party report libraries

In this section, we will use the third-party reporting library *Cluecumber* (https://github.com/trivago/cluecumber). This is an open source project that is mainly used for turning Cucumber JSON into test reports. Since Karate can write the same JSON format as Cucumber, we can use any other reporting library that is Cucumber compatible.

> **Karate native versus JSON-dependent reports**
> Contrary to the native Karate reports, which are generated by the framework itself while the tests are running, JSON-dependent reports need to run after all the tests are completed.

Setting the appropriate Karate options

By default, Karate does not produce JSON files of its test runs. It also creates report files, which we need to suppress since we want to plug in our own reporting library:

```
@Test
void testParallel() {
    Results = Runner.path("classpath:reporting")
        .outputCucumberJson(true)
        .reportDir("target/json-files")
        .outputHtmlReport(false)
        .parallel(2);
    assertEquals(0, results.getFailCount(), results.
getErrorMessages());
}
```

Let's look at these settings in more detail:

- The `outputCucumberJson(true)` setting tells Karate that it should write Cucumber-compatible JSON files that can be consumed by third-party reports.

- `reportDir` is optional and can be set to any folder that should include the generated JSON files. If this is omitted, they are written to `target/karate-reports` instead.

- The `outputHtmlReport(false)` setting suppresses Karate's built-in reports.

Next, we will add Cluecumber to our project dependencies.

Using Cluecumber as a Maven plugin

In our first example, we will use the Cluecumber plugin directly from Maven. This is rather straightforward since our tests are triggered via Maven as well.

Technical requirements

The approach we will be taking is demonstrated in an example project that can be found at `https://github.com/PacktPublishing/Writing-API-Tests-with-Karate/tree/main/chapter05/karate-reports-3rd-party-maven/`.

This project includes two feature files with five tests in total. I deliberately made a test fail to show how this looks in the newly integrated report. For visibility, I also put `@passing` and `@failing` tags on the respective scenarios.

Using the reporting dependency

If we want to use Cluecumber as a Maven plugin, we must add it to the `plugins` section inside the `<build>` block of our project's `pom.xml` file:

```
<build>
    <plugins>
        <plugin>
            <groupId>com.trivago.rta</groupId>
            <artifactId>cluecumber-maven</artifactId>
            <version>3.0.0</version>
        </plugin>

        ...
    </plugins>

    ...
</build>
```

Now, we need to add two mandatory settings to point this plugin to the generated JSON files and define the output folder. For this, we can add a `<configuration>` block inside the `<plugin>` block:

```
<configuration>
    <sourceJsonReportDirectory>target/myReport
    </sourceJsonReportDirectory>
    <generatedHtmlReportDirectory>target/customReport
    </generatedHtmlReportDirectory>
</configuration>
```

Let's look at this in more detail:

- `sourceJsonReportDirectory` points to the directory where the JSON files are stored

- `generatedHtmlReportDirectory` specifies the directory in which the generated report should be saved

Many more properties can be set for Cluecumber, such as customization options, display preferences, how it should deal with skipped tests, and so on. A full list can be found in the project's README file: `https://github.com/trivago/cluecumber-report-plugin/blob/main/README.md`.

Testing the setup

In this step, we will check if our new reports work.

> **Running the reporting step**
>
> Since we are running Karate as unit tests, they will stop Maven execution if a test fails. So, if we wanted to include report generation within the regular `mvn test` run, we would have to implement workarounds to make sure reports are triggered in the end, regardless of the state (such as ignoring the test result or running these tests as integration tests). We will not do this here as this would go beyond the scope of this chapter. Instead, we will simply execute a separate Maven command to trigger report generation after the test run – this is also what is typically done in CI/CD pipelines when using separate Maven reporting libraries.

We can run the reporting by executing this Cluecumber-specific Maven command:

```
mvn cluecumber:reporting
```

Of course, this needs to be run when JSON files exist, so a test run with the appropriate runner option to write JSON files must be done before we run this command.

This should generate a report that can be found at `target/customReport/index.html`, as specified:

```
19:25:52.929 INFO: => Cluecumber Report: target/customReport/
index.html
```

Opening the `index.html` file in a browser should produce a full report, like this:

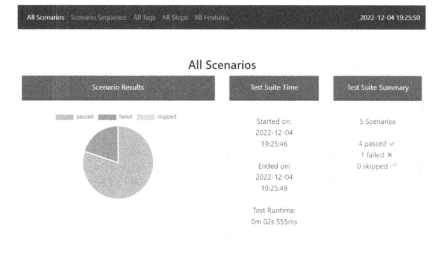

Figure 5.13 – Cluecumber report start page

> **Caution – pitfall!**
>
> It is important *not* to use `mvn clean cluecumber:reporting` here as this would wipe the target directory, including all generated JSON files. In this case, the generated report would be empty!

In the report, we can also see information about each scenario and the cause of failures. We can also click on the scenario's name to get all the details, such as the Karate report:

Figure 5.14 – Cluecumber report failure information

Now, let's check out the second way to use a third-party report directly from the code while using another Cluecumber library.

Using Cluecumber from a Karate runner

As we have seen, generating a report from Maven is not too hard but can be problematic when tests are failing. If a report must be generated from code, the approach is a little bit different but equally straightforward.

Technical requirements

This approach is demonstrated in an example project that is based on the Maven example mentioned previously. It can be found at `https://github.com/PacktPublishing/Writing-API-Tests-with-Karate/tree/main/chapter05/karate-reports-3rd-party-runner/`.

Using the reporting dependency

This time, we don't want to execute report generation through a Maven plugin. Instead, we want to do so directly in the runner code, so the Maven dependency is different:

```xml
<dependencies>
    ...
    <dependency>
        <groupId>com.trivago.rta</groupId>
        <artifactId>cluecumber-core</artifactId>
        <version>3.0.0</version>
    </dependency>
</dependencies>
```

The cluecumber-core dependency needs to be added as a <dependency> element within the already existing <dependencies> block. This is the same block where the Karate dependency has been added. Now, we can use it inside the runner class.

Modifying the runner code

Let's take a look at the runner code:

```java
@Test
void testParallel() throws CluecumberException{
    Results = Runner.path("classpath:reporting")
        .outputCucumberJson(true)
        .reportDir("target/myReport")
        .outputHtmlReport(false)
        .parallel(2);

    new CluecumberCore.Builder().build()
        .generateReports(
            "target/myReport",
            "target/customReport");

    assertEquals(0, results.getFailCount(),
        results.getErrorMessages());
}
```

Here, we can add the highlighted call to Cluecumber in between the code that runs the tests and the code that passes or fails the whole test run with the `assertEquals` call. This way, the report is always generated after the test run but before the build can fail, so the problem with the Maven plugin approach that requires two separate executions does not exist here.

Also, note that no configuration in the `pom.xml` file is necessary since we can configure the reporting options directly in the code.

From now on, running the test runner through VS Code or Maven produces the same report that we saw in the Cluecumber Maven example!

Finally, let's take a quick look at Karate's Surefire reporting options.

Generating a Surefire report

By default, Surefire, the Maven plugin running our unit tests, generates an XML summary that can be used in CI/CD pipeline projects or by specific report generators. Let's take a quick look at the format, how to control it, and how to generate a report from it via Maven.

Surefire XML

More files are written by Karate, and these can be found in the `target/surefire-reports` directory:

- The `reporting.Run.txt` file contains a summary of the test run, such as this:

```
-------------------------------------------------------
Test set: reporting.Run
-------------------------------------------------------
Tests run: 1, Failures: 1, Errors: 0, Skipped: 0, Time
elapsed: 5.782 s <<< FAILURE! - in reporting.Run
testParallel  Time elapsed: 5.751 s  <<< FAILURE!
org.opentest4j.AssertionFailedError:
status code was: 404, expected: 200, response time in
milliseconds: 261, url: https://jsonplaceholder.typicode.
com/this_is_wrong, response:
{}
classpath:reporting/demo/Reporting.feature:23 ==>
expected: <0> but was: <1>
      at reporting.Run.testParallel(Run.java:14)
```

  This is a condensed version of the last part of the usual command-line output.

- The `TEST-reporting.Run.xml` file includes a complete XML representation of the test run. This file can be used to generate Surefire reports.

Surefire XML schema

For more information about the contents of the Surefire XML format, you can look at the official XML schema definition (**XSL**) here: `https://maven.apache.org/surefire/maven-surefire-plugin/xsd/surefire-test-report-3.0.xsd`.

We will look at how we can use this format further in *Chapter 6, Karate in Docker and CI/CD Pipelines*. For now, let's use Maven to demonstrate what a report based on this format could look like.

Generating a report from Surefire XML

To generate a Surefire report, we can use Maven's `surefire-report` plugin. This is automatically triggered when we execute the following command in the Terminal:

```
mvn surefire-report:report
```

This will generate a new directory called `target/site` that contains the generated report.

When opening the `surefire-report.html` file in a web browser, we will see that there is indeed a report but without any CSS or images:

Last Published: 2022-12-06 | Version: 1.0-SNAPSHOT

Built by Maven

Surefire Report

Summary

[Summary] [Package List] [Test Cases]

Tests	Errors	Failures	Skipped	Success Rate	Time
1	0	1	0	0%	5.835

Note: failures are anticipated and checked for with assertions while errors are unanticipated.

Figure 5.15 – Surefire report without CSS and images

This is happening because this plugin can be configured with multiple skins to customize the report further and our used command just generates the basic content. Setting up the whole process is beyond the scope of this book.

What we can do, however, is use the default skin by running this command afterward:

```
mvn site -DgenerateReports=false
```

This tells Maven to just generate the missing CSS and image files and nothing more and add this to our `site` directory.

Now, we should see a slightly better-looking report:

Figure 5.16 – Full Surefire report

Like before, we can get information about the test suite and tests, their state, as well as their exception message and stack trace in case of failures:

Figure 5.17 – Surefire report failure display

The main difference is that we have everything together here on a single page. Therefore, reports like this are ideal for test overviews within build systems.

Finally, let's look at some other report files where Karate can generate JUnit XML.

Using JUnit reports

JUnit XML files are test-related files that can also be used by specialized build server plugins and other tools that can work with this format, such as xunit-viewer (https://lukejpreston.github.io/xunit-viewer/).

By default, JUnit XML files are not generated when running tests. Since we are dealing with a smart test framework, we can turn on this generation by using the Karate outputJunitXml(true) option in the Runner method:

```
Results = Runner.path("classpath:reporting")
    .outputCucumberJson(true)
    .reportDir("target/myReport")
    .outputHtmlReport(false)
    .outputJunitXml(true)
    .parallel(2);
```

This option adds another XML file for every test, as seen here:

Figure 5.18 – JUnit XML files in karate-reports

Each file contains all the information about the specific test runs. We can visualize this quickly by using a web tool such as the online JUnit parser at https://lotterfriends.github.io/online-junit-parser.

Pasting the contents of one of the generated XML files into it generates a quick overview, like this:

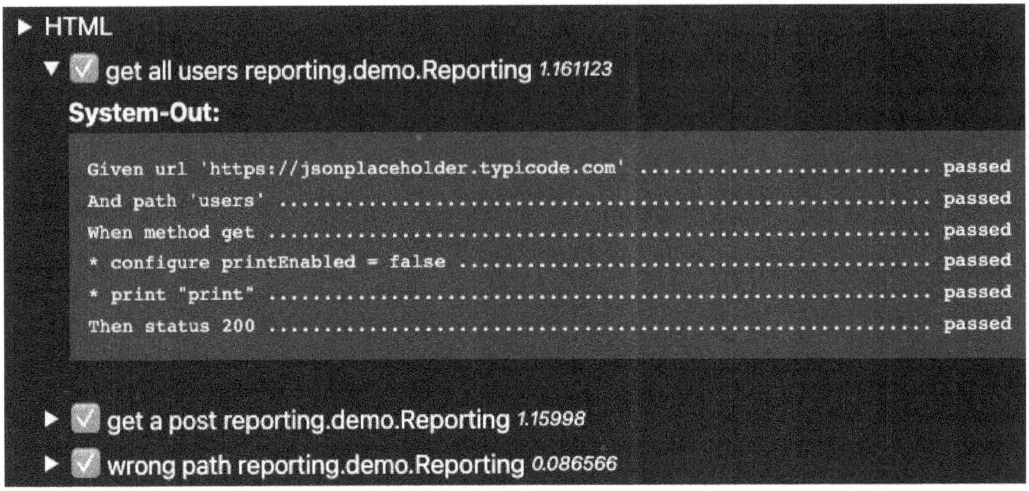

```
▶ HTML
   ▼ ✓ get all users reporting.demo.Reporting 1.161123
   System-Out:

   Given url 'https://jsonplaceholder.typicode.com' ........................ passed
   And path 'users' ........................................................ passed
   When method get ......................................................... passed
   * configure printEnabled = false ....................................... passed
   * print "print" ......................................................... passed
   Then status 200 ......................................................... passed


   ▶ ✓ get a post reporting.demo.Reporting 1.15998
   ▶ ✓ wrong path reporting.demo.Reporting 0.086566
```

Figure 5.19 – JUnit online viewer

As you can see, this format gives us information about our tests in a basic but still understandable form.

This concludes our overview of the logging and reporting options for Karate.

Summary

In this chapter, we looked at different ways to configure and use Karate's logging to make it fit our requirements.

After that, we looked at Karate's built-in report, what information it provides, and how we can customize it. Then, we looked at how to switch the default report to another reporting solution or use third-party solutions as an additional source of information. Finally, we took a quick look at the more specialized Surefire and JUnit XML files and how they can be used.

In the next chapter, we will delve into some of the more advanced Karate functions and concepts.

Part 2: Advanced Karate Functionalities

In this part, we will deepen and expand the knowledge from the first part. We will deal with headers and cookies, work with external files, and learn about JSONPath and XPath. After that, we will get to know different ways of realizing own functionality in Karate through JavaScript and Java. Karate tests are often used in build pipelines, so we'll look in detail at how that works. Lastly, we will go through browser testing with the Karate UI and performance testing with Karate Gatling to cover these specialized aspects as well.

This section contains the following chapters:

- *Chapter 6, More Advanced Karate Features*
- *Chapter 7, Customizing and Optimizing Karate Tests*
- *Chapter 8, Karate in Docker and CI/CD Pipelines*
- *Chapter 9, Karate UI for Browser Testing*
- *Chapter 10, Performance Testing with Karate Gatling*

6

More Advanced Karate Features

In the previous chapters, we looked at how to write and run Karate tests and how to use the various reporting options to make sense of the results. In this chapter, we will explore some more advanced concepts and functionalities so that we can solve more elaborate test cases.

In this chapter, we will cover these main topics:

- Working with headers and cookies
- Using different configuration and run options
- Defining and using expressions with `def`
- Working with external files
- Understanding JSONPath and XPath
- Testing GraphQL

Technical requirements

The code examples for this chapter can be found at `https://github.com/PacktPublishing/Writing-API-Tests-with-Karate/tree/main/chapter06`.

You will require the following:

- The system and IDE setup we completed in *Chapter 2, Setting Up Your Karate Project*
- Postman to explore our example GraphQL API

We will first look at how to work with headers and cookies—two of the building blocks of HTTP requests.

Working with headers and cookies

Headers and cookies are often used in web and API development to improve the user experience and functionality. Headers are pieces of information that are sent between the client and the server when a web page is requested. They can be used to specify the type of content that is being sent, the location of the content, and other details that can help the server process the request and send an appropriate response. Cookies, on the other hand, are small pieces of data that are stored on the client's computer and can be used to remember information about the user and their preferences. This information can be used by the website to provide a more personalized experience for the user, such as remembering their login details or preferences for the site. In this section, we will work with both.

The code examples for this section can be found in the project repository under `https://github.com/PacktPublishing/Writing-API-Tests-with-Karate/tree/main/chapter06/headers_cookies`.

More about headers

Headers are part of a request or response. In fact, we saw these before in *Chapter 3* when we made our first requests to the JSONPlaceholder API:

```
16:47:25.928 [com.intuit.karate.cli.Main.main()] DEBUG com.
intuit.karate - request:
1 > GET https://jsonplaceholder.typicode.com/posts?userId=1
1 > Host: jsonplaceholder.typicode.com
1 > Connection: Keep-Alive
1 > User-Agent: Apache-HttpClient/4.5.13 (Java/17.0.4.1)
1 > Accept-Encoding: gzip,deflate
```

Headers contain key-value pairs, as seen here. They are typically containing metadata about a request or response. This means it is data that is not part of the request payload or response data directly but data that describes the format, connection, client, and server.

Another important task of headers is to pass API keys or authentication tokens, mostly for non-public and commercial APIs, as we will see next.

Setting headers

For many public APIs, you need some form of authentication sent within the request headers. For this example, we will use yet another mocked API that acts like it has a typical authentication flow.

The API we will use for this example is `https://www.instantwebtools.net/secured-fake-rest-api`. Its endpoint can only be accessed with a valid authentication token that has to be requested from a specific authentication endpoint.

The flow of our example is as follows:

1.   Request an authentication token for our API user.

2. Query an endpoint using the requested token in the request header.

3. Assert that our request was successful.

Luckily, this API is well documented and even provides us with valid usernames and passwords that we can use.

Exploring the API under test

To illustrate the flow, let's look at the API in Postman:

Figure 6.1 – Requesting the OAuth token

Here, we send the different keys and values specified in the documentation to the token endpoint of the authentication API. Since this has to be sent in the **x-www-form-urlencoded** format, we need to select this option here as well:

Figure 6.2 – Returned token in the response body

Sending this request returns the `access_token` value in the response body that we can then use to authenticate against the next API:

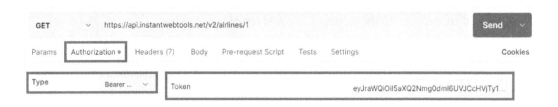

Figure 6.3 – Passing the token as an authorization header in Postman

When calling the `airlines` endpoint, we need to authenticate with the retrieved token. The type must be **Bearer**, and the full token string we received earlier has to be copied to the **Token** field.

> **Note:**
> Be careful to copy the value without the enclosing quotes! This token is only valid for a limited time, so this must be done regularly.

Implementing the Karate scenario

Let's now create the example as a Karate scenario.

Retrieving the token

First, we need to retrieve the token to pass as an auth header to the final API. For this, we need to pass some parameters to a special authentication API that issues such a token based on valid credentials. Luckily, our API documentation specifies exactly what these parameters must be. In a real-world scenario, you would have to register with an API provider first to receive these credentials:

```
Background:
  Given url 'https://dev-457931.okta.com/oauth2/
aushd4c95QtFHsfWt4x6/v1/token'
  * def credentials = { scope: 'offline_access', grant_
type: 'password', username: 'api-user4@iwt.net', password:
'b3z0nV0cLO', client_id: 'OoahdhjkutaGcIK2M4x6' }
  And form fields credentials
  When method post
  * def token = response.access_token
```

In this case, the steps necessary to retrieve the token are in a `Background` scenario since they might be relevant to multiple tests. However, in *Chapter 7, Customizing and Optimizing Karate Tests*, we will see how this can be solved even more elegantly.

First, we put all needed key-value pairs into a JSON structure and initialize the `credentials` variable with this. The documentation told us to send these parameters as form fields so that we can use Karate's `form fields` keyword to achieve this.

In the returned JSON structure, the `access_token` property has the value we need to pass in the header of the next API call.

Using the token

The `token` variable from the `Background` scenario is available to the other scenarios so that we can use it here as well. As in our initial Postman exploration, we query `/airlines/1` (split into `url` and `path` in Karate). The key is to set the `Authorization` header and give it a value of `'Bearer ' + token`. This is then sent with our GET request, unlocking the API for us to use:

```
Scenario: authenticate and get airplane data
  Given url 'https://api.instantwebtools.net/v2/airlines'
  And path '1'
  And header Authorization = 'Bearer ' + token
  When method get
  Then status 200
  * match response.name == "Quatar Airways"
```

Without a valid token, these APIs typically return a `401 (Unauthorized)` error.

Checking response headers

Headers are sent with requests and responses. If you want to know what the headers include, there is the `responseHeaders` property. This one contains a JSON structure of all keys and values that are included. For our example, using `* print responseHeaders`, it looks something like this:

```
10:39:58.289 [pool-1-thread-1] INFO  com.intuit.karate -
[print] {
  "Content-Type": [
    "application/json; charset=utf-8"
  ],
  "Content-Length": [
    "307"
  ],
  ...
}
```

For asserting values from the header, it is important to note that each value is an array, even if there is just one single piece of data.

Let's check that the Content-Length header has the correct value by comparing it to the actual size of the returned body. For this, we need to add the following line to our scenario:

```
* match responseHeaders['Content-Length'][0] == responseBytes.
length
```

Here, we compare the value stored in the response headers with Karate's responseBytes, a special variable that contains the response body in bytes. The length property returns the actual size, so we can directly compare it to the header value.

However, the data types won't match like this since the responseHeaders array contains all values within the arrays as strings, whereas responseBytes contains the size as an integer. To quickly circumvent that, we need to convert one side of the equation to the data type of the other. The easiest way, in this case, is comparing the responseHeaders value to '' + responseBytes.length:

```
* match responseHeaders['Content-Length'][0] == '' +
responseBytes.length
```

This automatically turns the numeric value into a string that can then be matched correctly. Another way to solve it would have been to convert the string value of the left side using the parseInt function, like this:

```
* match parseInt(responseHeaders['Content-Length'][0]) ==
responseBytes.length
```

This shows yet again how convenient the built-in JavaScript capabilities of Karate are.

More about cookies

Usually, API requests are stateless, meaning that each request is not linked to the one before. Cookies are small, specialized pieces of header data (normally as key-value pairs) that are mainly used for session management in an API context. Responses send these back, and the client can reuse them for subsequent requests to basically tie them together into one session. In a web browser, they are typically stored in and read from small text files that the browser saves locally.

Working with cookies

Using cookies is not that much different from the other headers. You can check the cookies that are sent back from the server with responseCookies. This is a JSON structure that can be used to make assertions of specific cookies in the response.

The nice thing about Karate is that you won't have to set response cookies on the next request—Karate will do this automatically. In case you need to set custom cookies, though, there is a simple way to do this.

```
Scenario: Cookies
  Given url 'https://api.instantwebtools.net/v2/airlines'
  And cookies { value1: 'test', value2: '123'}
  When method get
  * print karate.prevRequest.headers.Cookie
```

If the debug logs are on, you can see the cookies right in the header of the request:

```
22:38:06.153 [pool-1-thread-1] DEBUG com.intuit.karate -
request:
1 > GET https://jsonplaceholder.typicode.com/todos/1
1 > Cookie: value2=123; value1=test
```

You can also use the special `karate.prevRequest` variable to access details of the just-sent request. With `karate.prevRequest.headers.Cookie`, you can see the cookie values as well.

Hopefully, you have seen in this chapter that working with headers and cookies is not too complex in Karate. Let's look at more options to execute tests next, as there is more to discover here!

Using different configuration and run options

In *Chapter 4, Running Karate Tests*, we already discovered a lot of ways to trigger and execute tests both from the IDE and from the command line via Maven. Karate itself has some more interesting options that can be put to good use to make sure we are running tests exactly how we want. In the next sections, we will explore some of the most important ones.

Using the karate object for configuration and execution

We have already come across the `karate` object in the earlier chapters, most notably in *Chapter 5, Reporting and Logging*. Here, we used it to apply a few reporting options to all scenarios directly from `karate-config.js`. This was done using the `karate.configure` method with specific keys and values. An example was the option to suppress print statements in the logs:

```
karate.configure("printEnabled", false);
```

Also, we came across the `karate` object earlier when doing things such as setting the environment with `karate.env` and logging messages with `karate.log`.

> **A note about the karate object**
>
> All JavaScript and Karate expressions (these are the ones starting with `def`—we will explore these later in this chapter) can use various methods in the `karate` object. This is like an excessive utility class that provides functionalities for very different use cases. The scope of this book makes it impossible to discuss all manifestations of it. Therefore, we limit this to a few useful examples to illustrate how it works. Also, in later chapters, we will come back to the `karate` object again where it makes sense.
>
> For a complete list of all functionalities, it is worth looking at the official Karate documentation at `https://github.com/karatelabs/karate#the-karate-object`.

In this section, we will look at some more interesting options of the `karate` object concerning the execution and configuration of tests.

Selecting tests based on the operating system

Sometimes it might be necessary to run tests only on specific operating systems, especially if the API under test behaves differently based on the client OS or if certain error cases only happen on a specific platform. Here, Karate helps us with the `karate.os` property that contains information about the platform.

Let's write a test scenario that is only executed on Windows:

```
Scenario: Run only on Windows with abort
    * print karate.os
    * def isWindows = karate.os.type == "windows"
    * if (!isWindows) karate.abort()
    * print "I am running on Windows!"
```

If we run this scenario on Windows, the step printing `karate.os` gives us an output along these lines:

```
18:50:16.210 [pool-1-thread-1] INFO  com.intuit.karate -
[print] {
  "name": "Windows 11",
  "type": "windows"
}
```

Karate returns us the specific operating system name and version as the `name` key in the resulting JSON. In our case, the `type` key is more interesting. This contains a more generic name, so we can set the `isWindows` variable to `true` only if the type equals `windows`.

Next, we use an `if` condition based on the `isWindows` variable. If it is `false`, `karate.abort()` cancels the test and does not execute any following steps. This makes sure that the rest of the scenario is only executed on Windows.

Failing and aborting tests based on custom conditions

We have seen the `karate.abort()` method in the previous example. This just stops execution and skips all subsequent tests, making the scenario pass. It might be necessary in some cases to fail the test in a case such as this. This could be required if tests use operating system-specific functionality.

Let's quickly change the previous example and add a `failure` condition:

```
Scenario: Run only on Windows with failure
    * def isWindows = karate.os.type == "windows"
    * if (!isWindows) karate.fail("This is not windows!")
    * print "I am running on Windows!"
```

We could also do a similar thing with a matcher. However, using `karate.fail`, you can add a custom error message in this case. This will be visible in the logs:

```
>>> failed features:
This is not windows!
```

In the test report, it is also clearly visible:

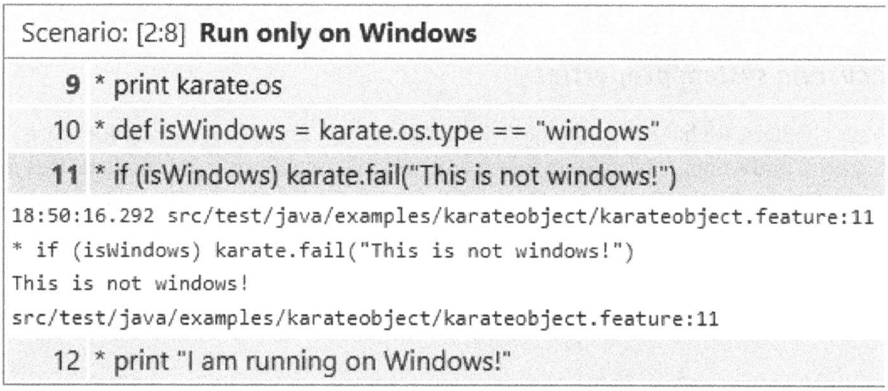

Scenario: [2:8] **Run only on Windows**

```
 9 * print karate.os
10 * def isWindows = karate.os.type == "windows"
11 * if (isWindows) karate.fail("This is not windows!")
18:50:16.292 src/test/java/examples/karateobject/karateobject.feature:11
* if (isWindows) karate.fail("This is not windows!")
This is not windows!
src/test/java/examples/karateobject/karateobject.feature:11
12 * print "I am running on Windows!"
```

Figure 6.4 – karate.fail and skipped step

We can see the `* if (!isWindows) karate.fail("This is not windows!")` failed step in *line 11*. The next one in *line 12* is skipped. It is marked in a different color and is not expandable like the others because it naturally does not have any log output.

Calling operating system commands

Another operating system-specific feature that can be useful in tests that involve not only an API but also the command line can be Karate's ability to call system commands. This scenario uses `karate.exec` to determine the Java version information by calling the `java` command with the `--version` flag. This is passed to the operating system to execute, and after it is finished, Karate continues with the next test steps:

```
Scenario: Get Java information
  * def javaInfo = karate.exec("java --version")
  * print javaInfo
```

This prints out something like this:

```
19:20:53.497 [pool-1-thread-1] INFO  com.intuit.karate -
[print] java 17.0.4.1 2022-08-18 LTSJava(TM) SE Runtime
Environment (build 17.0.4.1+1-LTS-2)Java HotSpot(TM) 64-Bit
Server VM (build 17.0.4.1+1-LTS-2, mixed mode, sharing)
```

Next, we'll see how to pass custom system properties.

> **Non-blocking execution**
>
> If you need to call system commands that should not block Karate's test execution, you can use `karate.fork`, which executes commands in another parallel thread.

Passing custom system properties

In the former chapters, we have used the built-in Karate system properties such as `karate.env`, `karate.tags`, and so on. It is possible to use any custom system properties as well. This makes it possible to pass values such as `baseUrl` from the outside without being dependent on specific `karate.env` values.

System properties that are not wired into Karate can be accessed using the `karate.properties` array within the `karate-config.js` file, like this:

```
function fn() {
  var config = {
    name: karate.properties["name"]
  }
  return config;
}
```

This sets the name key to the value of the system property name. As with any other config variables, this is now available to all tests:

```
Scenario: Using system properties
    * print "Hi, my name is", name
```

As we have already discussed previously, there are multiple ways to set these values. Let's check quickly that this works as expected, using the `systemProperty` runner method:

```
@Karate.Test
Karate testProperty() {
    return Karate.run("properties")
        .systemProperty("name", "Jane")
        .relativeTo(getClass());
}
```

Running this method now, our test prints the value as expected:

```
19:34:57.992 [main] INFO  com.intuit.karate - [print] Hi, my
name is Jane
```

As with the built-in Karate properties, these can also be set via the Maven command line using -Dname=Jane from the outside.

Inside a parallel runner, you can use Java's `System.setProperty` method:

```
@Test
void testParallel() {
    System.setProperty("name", "Richard");
    Results results = Runner.path("classpath:examples")
            .parallel(5);
    assertEquals(0, results.getFailCount(),
            results.getErrorMessages());
}
```

As you can see, these custom system properties add another layer of flexibility to the already powerful Karate properties.

Request retries

A very useful Karate feature is a built-in request retry using the `retry until` keyword. This can be a lifesaver when dealing with testing workflows including external APIs that are flaky. Also, it allows us to test use cases where the polling of an API is needed until a certain result is reached.

For unreliable APIs, this approach is very common:

```
Scenario: Retry based on status
    Given url 'https://jsonplaceholder.typicode.com/todos/1'
    And retry until responseStatus == 200
    When method get
```

Here, we can use Karate's `responseStatus` variable to wait until it returns `200 (OK)`. If this is not the case, Karate will retry it three times with an interval of `3000` milliseconds between each try. This is Karate's default retry behavior.

Note that the number of retries *includes the initial call*, so a retry value of 3 means *try it two more times after the failed first call*!

This is also reflected in the logs:

```
22:06:02.421 [main] DEBUG com.intuit.karate - retry condition
not satisfied: responseStatus == 200
22:06:02.421 [main] DEBUG com.intuit.karate - sleeping before
retry #1
```

If this condition is met within the allowed retries, the test continues successfully. Otherwise, the test fails, and this line is visible in the logs:

```
too many retry attempts: 3
```

It is important to note, though, that the `retry until` condition has to be written in pure JavaScript as Karate internal mechanisms such as `match` or `assert` will not work here.

There are a couple of ways to tweak the default retry behavior further, as we will see next.

Configuring retries in karate-config.js

We can configure the retry behavior right inside `karate-config.js` to make it available across the whole test suite:

```
karate.configure('retry', { count:2, interval:1000 });
```

This example configuration sets the retry count to 2 with an interval of 1 second between each retry.

Configuring retries in test code

We can also configure the retries directly within the feature files to customize the behavior for specific test cases. This can be done with the `configure` keyword that we already know from previous chapters:

```
Scenario: Retry based on custom condition
    Given url 'https://jsonplaceholder.typicode.com/todos/1'
    * configure retry = { count: 2, interval: 5000 }
    And retry until response.userId == 1 && response.name ==
"test"
    When method get
```

Here, the same `retry` key pointing to a JSON object with `count` and `interval` keys is used in conjunction with the `configure` keyword. This setting overwrites the default settings and the settings coming from `karate-config.js`.

The retry configuration is active for all following method calls, so it must be placed before the actual method call it should be used for!

Using advanced tags

We have seen the power of tagged scenarios in *Chapter 4, Running Karate Tests*. Let's now look at two special kinds of tags that are very useful when running specific test suites.

Running environment-specific scenarios using tags

In *Chapter 4*, we have seen how we can select different environments and adapt the configurations and endpoints accordingly. Sometimes, it might be useful to have specific scenarios that only run in one environment but not in others. An example is some developer tests that should only run while developing a new feature. Another is a specific smoke test suite that is only relevant for production use.

Let's take this feature file as an example:

```
Feature: Using tags
    @env=dev
    Scenario: Dev only
        * print "DEV ONLY"

    @env=dev,prod
    Scenario: Dev and prod
        * print "DEV and PROD"
```

```
@envnot=dev
Scenario: Not in dev
    * print "Not in DEV"
```

We have two scenarios that have the special Karate @env tag that can specify after = which scenario or scenarios it should run in. If you want to give it multiple scenarios, they should be separated by commas. Please note that there should not be any spaces in the tags at all.

The last scenario is marked with an @envnot tag. This is the opposite of @env and tells Karate that these tests should *not* run in one or more specific environments.

If we now run these scenarios with karate.env set to dev, we get this output:

```
22:16:01.244 [main] INFO  com.intuit.karate - [print] DEV ONLY
22:16:01.287 [main] INFO  com.intuit.karate - [print] DEV and
PROD
```

So, indeed, both scenarios containing dev as a value of the @env tag are picked up by Karate, whereas the one having dev in @envnot is ignored.

Running with karate.env set to prod, we correctly get this:

```
22:29:36.470 [main] INFO  com.intuit.karate - [print] DEV and
PROD
22:29:36.507 [main] INFO  com.intuit.karate - [print] Not in
DEV
```

Only the scenario for prod and the one with dev in the @envnot tag are correctly run this time!

Using value tags

You might notice that the environment tags are special in two ways:

1. They are built into Karate
2. They are composed of the tag plus some values

The latter can be used with custom tags as well, giving you even more control over what to run and how.

This feature shows the power of this approach:

```
Feature: Value tags
    @car=audi
    Scenario: Value tags
        * print karate.tagValues
```

```
@car=mercedes,tesla
Scenario: Value tags
    * print karate.tagValues
```

The `karate.tagValues` variable stores the scenario tags as a convenient JSON structure, so we could even use these within a scenario or in custom methods. Running the second scenario, for example, would print out this:

```
22:47:35.098 [pool-1-thread-2] INFO  com.intuit.karate -
[print] {
  "car": [
    "mercedes",
    "tesla"
  ]
}
```

We can use these tags through the normal tag mechanism of Karate—for example, with a runner method:

```
@Karate.Test
Karate testValueTags() {
    return Karate.run("value_tags")
        .tags("@car=audi").relativeTo(getClass());
}
```

This runs only the first scenario as it is tagged with `@car=audi`. Combining these value tags with the other mechanisms for picking specific tests and test suites to run makes it possible to respond completely flexibly to all possible test requirements within the **Software Development Life Cycle** (**SDLC**).

Now that we have seen yet some more interesting execution options of the Karate framework itself, it is time to dive deeper into the power of the `def` command.

Defining and using expressions with def

In earlier chapters, we used the `def` keyword to define variables (for example, `* def myName = "Benjamin"`). However, it can be way more flexible and time-saving than that if we use it for custom functionality as well. In this section, we will explore some of these aspects.

Defining inline methods

The def keyword allows us to define helper functions easily so that we don't have to repeat the same calculations, string manipulations, and so on repeatedly. Here's an example:

```
Scenario: Miles and kilometers
    * def kmToMiles = function(km) { return km / 1.6 }
    * def milesToKm = function(miles) { return miles * 1.6 }
    * assert kmToMiles(16) == 10
    * def miles = kmToMiles(90)
    * match miles == 56.25
    * match milesToKm(miles) == 90
```

In this example, there are two functions: kmToMiles, which takes a km parameter and returns the number of miles, and the milesToKm function, taking in a miles parameter that does the reverse calculation. These functions are defined as the right side of the expressions and do not have a name (hence they are called *anonymous* functions). That means that we can assign them to any variable we want.

To use these, we just need to specify the variable name and add parentheses plus the required parameters if there are any (for example, kmToMiles(16)). The return values of these functions can be used in assert and match statements, or they can be assigned to new variables and reused in later steps.

Also, the methods themselves can be assigned to new variables:

```
    * def someOtherFunctionName = kmToMiles
    * match 2 == someOtherFunctionName(3.2)
```

Here, the kmToMiles variable that contains a function is assigned to the someOtherFunctionName variable that can then be used like the original function by calling it like this: someOtherFunctionName(3.2). This can be useful when functions are nested, or you want to increase clarity by renaming a function to match a specific context.

Using embedded expressions with JSON

There can be use cases where you need to have a larger JSON or XML file that you need to use as a request payload but that should be partly dynamic. Karate supports such a templating approach. This means that you can define the outline of a JSON or XML file and define placeholders that can be set dynamically. These templates can either be defined within a scenario or read from an external file (we will explore the different file options in the next section of this chapter).

This example shows the power of this approach. It uses Karate expressions to construct a JSON object that represents a book.

All keys are given here, but all values come from variables and functions:

```
Scenario: Using JSON templates
    * def title = 'Testing APIs'
    * def pages = 250
    * def chapters = ['Core concepts', 'Setup']
    * def titleBackwards = function(title) { return title.
split("").reverse().join(""); }
    * def format = null
    * def book =
    """
      {
        title: '#(title)',
        pages: #(pages),
        chapters: #(chapters),
        totalChapters: #(chapters.length),
        totalChaptersPlusIntro: #(chapters.length + 1),
        titleBackwards: '#(titleBackwards(title))',
        format: '##(format)'
      }
    """
    * print book
```

There is quite a lot to uncover here because this feature is quite powerful.

The general approach to using templating in Karate is the # () expression syntax. Everything that is between the parentheses is parsed and executed, and expressions must always start with # (and end with). This way, variables and functions can be used here to replace these expressions.

Also, it is important to enclose expressions within quotes when they are strings. Otherwise, the resulting JSON will be invalid. An example of such a case is the '# (title) ' expression in our book JSON.

In the case of the chapters: # (chapters) line in our JSON template, it is important to note that you cannot just replace an expression with a string or numeric value but also with more complex structures and even full JSON objects. In our case, an array of strings representing chapter titles is injected here.

This approach is not limited to variables, but you can use pretty much all JavaScript functionality in here, even externally defined functions. The totalChapters: # (chapters.length) line shows that we can use the array length of our injected chapters array directly without any custom hacks—this represents the total number of chapters in the book (in our example case, it is 2).

`totalChaptersPlusIntro: #(chapters.length + 1)` demonstrates that you can even make calculations right within an expression. To take this even further, `titleBackwards: '#(titleBackwards(title))'` uses a custom function that is defined in the `titleBackwards` variable. This can then be used again right in the expression.

The last element, `format`, is special. Here, we use Karate's `##` optional marker to remove this element completely if it is `null`. In this case, there will be no `format` key and value in the resulting JSON as the `format` variable is deliberately set to `null` here.

Here is the complete resulting JSON array with the replaced expressions highlighted:

```
20:06:26.323 [pool-1-thread-1] INFO  com.intuit.karate -
[print] {
  "title": "Testing APIs",
  "pages": 250,
  "chapters": [
    "Core concepts",
    "Setup"
  ],
  "totalChapters": 2,
  "totalChaptersPlusIntro": 3,
  "titleBackwards": "sIPA gnitseT"
}
```

This approach is very flexible and concise and can be adapted for a wide variety of use cases.

> **Removing JSON elements**
>
> We have seen how to set optional template elements by setting them to `null`. This makes it possible to include optional elements based on certain conditions.
>
> If you want to omit elements from already *existing* JSON, Karate offers the `remove` keyword to do just that (for example, `* remove book.title`).

Let's look at a small example of embedded expressions in XML.

Using embedded expressions with XML

Luckily, embedded expressions can be used almost like in JSON. This is a shortened version of the data we used in the JSON example:

```
Scenario: Using XML templates
  * def title = 'Testing APIs'
```

```
* def pages = 250
* def book =
"""
  <book>
    <title>#(title)</title>
    <pages>#(pages)</pages>
  </book>
"""
* def book = book.book
* match book.title == title
* match parseInt(book.pages) == pages
```

The only difference here, apart from the data format of the template, is the added line def book = book.book, which redefines the book variable from the beginning to point to the XML elements enclosed by the <book></book> tag so that we can directly use the inside elements for our matches.

You can see that templating and embedded expressions are very powerful features in Karate that can greatly reduce the complexity and increase the readability of test cases.

As we can also take these templates from external files, let's explore next how Karate handles read operations with different file types.

Working with external files

We have already discussed the data types that Karate can natively work with. Now, we will see that this also works when reading these from external files. This is a great feature since it allows us to keep the tests small and concise even when working with large amounts of data. This would not be the case if we put this data right into the test scenarios.

There is, of course, also a downside when working with files as opposed to inline data—our tests and the data that they need are separated from each other and therefore require more steps if someone wants to follow the flow of the test. So, you must think carefully about what makes more sense in each individual case.

The code examples for this section can be found in the project under https://github.com/PacktPublishing/Writing-API-Tests-with-Karate/tree/main/chapter06/files. The project setup is shown here:

Figure 6.5 – File examples project setup

In here, you can find examples of all the different file types that we will discuss in the following sections. All files are located directly in the `examples` directory, and `files.feature` inside the `files` directory contains scenarios to read them. The `.json`, `.xml`, and `.csv` files store the name `Karate` and its inception year `2017` in the respective format.

The `.txt` and `.js` files, on the other hand, are processed differently, as their main purpose is not to store structured data, but rather free-flowing content and additional functionality, respectively.

Working with JSON files

For this example, the information is stored in a file called `json-example.json`, which looks like this:

```
{
    "name": "Karate",
    "inception": 2017
}
```

We can work with this in a test scenario by using Karate's `read` keyword:

```
Scenario: Read JSON
  * def json = read('../json-example.json')
  * assert json.name == "Karate"
  * assert json.inception == 2017
```

The file contents are directly stored in native JSON format in the json variable so that we can directly assert the value of the two keys, name and inception.

It is also possible to read JSON from a file and then make changes to it as if it were directly defined within the scenario itself!

Working with XML files as JSON

Let's express the same information as before in XML form as contained in the xml-example.xml project file:

```
<framework>
    <name>Karate</name>
    <inception>2017</inception>
</framework>
```

This is a very simple way of doing this. In this case, we can use it almost like in the previous JSON example. The only difference here is that we also need to read the root framework tag:

```
Scenario: Read XML
    * def xml = read('../xml-example.xml')
    * match xml.framework.name == "Karate"
    * assert xml.framework.inception == 2017
```

In the *Understanding JSONPath and XPath* section later in this chapter, we will see how we can work with more complex XML.

Reading CSV files

Comma-separated values (CSV) files typically express a two-dimensional table-like structure. So, we can convert the Karate information into this (as seen in the csv-example.csv file):

```
name,inception
Karate,2017
```

Here, we basically have one row in a table with two columns: name and inception. Karate will convert this into a JSON structure automatically when reading it, like so:

```
Scenario: Read CSV
    * def csv = read("../csv-example.csv")
    * assert csv[0].name == "Karate"
    * assert csv[0].inception == 2017
```

Again, the assertions look very close to the ones before.

It is noteworthy that we need to specify the index of the row when reading the values since this structure is represented as a JSON array internally. We can verify this by adding a `* print csv` line after reading from the file:

```
08:55:03.823 [main] INFO  com.intuit.karate - [print] [
  {
    "name": "Karate",
    "inception": "2017"
  }
]
```

Let's move on to two different file formats: text and JavaScript.

Reading YAML files

We have a simple YAML file called `yaml-example.yml` that contains the following:

```
framework:
  name: Karate
  inception: 2017
```

Here again, we can access the properties directly as Karate converts it to a JSON structure:

```
Scenario: Read YAML
  * def yaml = read('../yaml-example.yaml')
  * assert yaml.framework.name == "Karate"
  * assert yaml.framework.inception == 2017
```

Of course, as in all other examples before, an alternative way to do the value checks would be with a `match` statement—for example, like this:

```
* match yaml == {'framework': {'name': 'Karate', 'inception':
2017}}
```

This too works because of the smart JSON conversion!

Converting YAML and CSV to JSON

We have seen that Karate automatically converts YAML and CSV files to JSON to work with these more efficiently. A similar conversion is also available when creating these formats as strings, but we must explicitly do that as it is not done automatically.

Here is an example of YAML conversion:

```
Scenario: Convert YAML to JSON
  * text yaml =
    """
    publisher: Packt
    book:
      title: API testing with the Karate framework
    """
  * yaml converted = yaml
  * print converted
```

There are multiple interesting aspects in this example. First, we use the text command to tell Karate that the following string that is stored in the yaml variable is a plain text. Here, we must be careful to adhere to the YAML formatting so that the later conversion works.

> **Here document**
>
> You probably saw that here we use three double quotes (" " ") at the beginning and end of a string. This notation is typically called a **here document**; in Karate, it is named a **multiline expression**. This is a special way to define a formatted text because line breaks and whitespaces are taken exactly as they are written between these quotes. This is especially important in formats that are whitespace dependent, such as YAML, or in general for making data more readable.

If we now tried to do any JSON operation or matching on this yaml variable, it would not work since it is considered a text. So, to convert it to our desired JSON format, we use the yaml keyword. This means that the text should be treated as YAML, which is then finally turned into JSON:

```
23:18:09.102 [main] INFO  com.intuit.karate - [print] {
  "publisher": "Packt",
  "book": {
    "title": "API testing with the Karate framework"
  }
}
```

For CSV, this is almost the same:

```
Scenario: Convert CSV to JSON
  * text csv =
    """
    publisher,book
```

```
    Packt,API testing with the Karate framework
    """
* csv converted = csv
* print converted
```

The difference is that the `csv` keyword is used here.

This leads to the following JSON file:

```
22:55:13.521 [main] INFO  com.intuit.karate - [print] [
  {
    "publisher": "Packt",
    "book": "API testing with the Karate framework"
  }
]
```

Again, as seen before, this is turned into an array of JSON objects because each line in a CSV file is a separate data row. Also, for the automatic conversion to work correctly, the CSV format needs to start with a line that defines the keys of each column (here, it is `publisher` and `book`).

Reading text files

Naturally, a text file can hold pretty much all kinds of information and does not have a specific structure. So, these files cannot be turned into JSON automatically and they stay as strings.

Our text file includes the following sentence:

```
Karate had its first release in 2017.
```

We can read this file using a similar approach to before:

```
Scenario: Read text
  * def text = read("../text-example.txt")
  * assert text == "Karate had its first release in 2017."
```

Since this is just a string, we can only assert the text itself here.

Reading JavaScript files

JavaScript files are special in Karate since they can be used to store functionality that can be used within the test scenarios.

This is an example of a file called `js-example.js` containing the same valid JavaScript code that we previously used inside of a scenario:

```
function getRelease(name, year) {
    return name + " had its first release in "
        + year + ".";
}
```

This `sayHello` function takes two parameters (`name` and `year`) and returns a string containing these two parameters that resembles the contents of the text file from before (for example, for `Karate` and `2017`, it returns `Karate had its first release in 2017.`).

We can now read this file and store it in `jsFunctionFromFile`:

```
Scenario: Read Javascript
    * def jsFunctionFromFile = read("../js-example.js")
    * def text = jsFunctionFromFile("Karate", 2017)
    * assert text == "Karate had its first release in 2017."
```

The peculiarity now is that the function from the JavaScript file is not called by its original name (`sayHello`). Instead, Karate turns our `jsFunctionFromFile` variable from the test scenario into this function. Therefore, we need to call the function here as `jsFunctionFromFile("Karate", 2017)`.

After that, as with the text file before, we can use the returned string in our assert.

Defining multiple functions in one JavaScript file

Since Karate takes the first function in the JavaScript file and allocates it to the variable, you cannot access multiple functions from the same file in this way unless you create a wrapper function that includes all functions you need, as follows:

```
function() {
    return {
        getRelease: function(name, year){
            return name + " had its first release in "
                + year + "."
        },
        secondMethod: function(){ return 123 }
    }
}
```

Here, we have an anonymous function without a name whose only purpose is to return an object. The keys of the object (`getRelease` and `secondMethod`) point to values that are also functions—again, they are anonymous.

The `getRelease` function accepts two parameters as before; the `secondMethod` function just returns a number for illustration purposes.

Now, we can use these within a test scenario:

```
Scenario: Read JavaScript multiple methods
    * def f = read("../js-example2.js")
    * def text = f().getRelease('Karate', 2017)
    * print text
    * def number = f().secondMethod()
    * print number
```

As before, we can read the `js-example2.js` file and store it in a variable; I called it `f` for *functions* here. Since invoking `f()` executes our wrapper function, this returns the object containing our child functions. Now, we can execute each function just by referring to it by the respective key—for example, `f().getRelease('Karate', 2017)`.

When executing this scenario, we receive the expected outputs:

```
11:27:50.245 [main] INFO  com.intuit.karate - [print] Karate
had its first release in 2017.
11:27:50.248 [main] INFO  com.intuit.karate - [print] 123
```

A similar outcome can be achieved by adding functions to `karate-config.js`, as we will see next.

Using functions from karate-config.js

Since the `karate-config.js` file is executed automatically before all test runs, it means that not only properties with values but also properties containing functions are available to the scenarios:

```
function fn() {
    config = {
        function1: function(){ return "Cool function 1" },
        function2: function(){ return "Cool function 2" }
    }
    return config;
}
```

In the usual `config` object that is returned by the standard configuration method, we can store these functions (here called `function1` and `function2`) and use them in our scenarios directly:

```
Scenario: Methods from karate-config.js
    * def f1 = function1()
```

```
* print f1
* print function2()
```

This produces the expected output:

```
23:54:42.210 [main] INFO  com.intuit.karate - [print] Cool
function 1
23:54:42.217 [main] INFO  com.intuit.karate - [print] Cool
function 2
```

Both approaches for storing and using the JavaScript function shown before work fine if you have only a few functions that you need to be available to your scenarios. For more functionality, this can get messy and hard to maintain quickly. Later, in *Chapter 7, Customizing and Optimizing Karate Tests*, we will revisit this to discover an even better way to accomplish this.

Understanding JSONPath and XPath

Up until now, we have always used the basic syntax to traverse JSON data and access certain elements and values within it. In this section, we will look at **JSONPath** and **XPath**, which offer more flexibility when dealing with JSON and XML respectively.

This is not meant as a reference but rather a demonstration of how these two approaches work. You are encouraged to learn more about these as they offer a lot of functionality in themselves!

Working with XPath

Let's look at XPath first as JSONPath is loosely based on it.

> **XPath details**
>
> XPath is a language to parse XML structures. It considers an XML document as a node tree of elements, attributes, and text and can find XML parts in children and parent nodes. More information about XPath can be found here:
>
> https://www.w3.org/TR/1999/REC-xpath-19991116/

We will use this XML structure within a Karate test scenario:

```
Scenario: Xpath
  * def data =
  """
  <magicians>
    <magician id="1">
```

```
        <name>Harry Houdini</name>
        <birthyear>1874</birthyear>
        <specialty>escapology</specialty>
        <specialty>card tricks</specialty>
      </magician>
      <magician id="2">
        <name>David Copperfield</name>
        <birthyear>1956</birthyear>
        <specialty>illusions</specialty>
      </magician>
    </magicians>
    """
```

This is a list of magicians with name, birthyear, and specialty tags. Also, each one has an id attribute. We can now traverse and match data using XPath directly in Karate! Here's an example:

```
* match data count(/magicians/magician) == 2
```

Here, we use a special match syntax that accepts an XPath expression. First, it will find all children under the magicians root tag that have a magician tag. In the second step, the count function, which is part of the XPath specification, returns the number of matching elements. We can directly use this in Karate to assert the expected number of elements.

We can also use more complex expressions including filters, like this:

```
* match data //magician[@id='2']/name == 'David Copperfield'
```

This expression finds all elements with magician tags no matter where they are (this is what //magician stands for) and filters by the id attribute with value 2 (@id='2'). Again, we can directly use this result in a match statement and compare it to the expected name.

As a sidenote, it is also possible to store this result in a variable and then use it afterward with Karate's get keyword:

```
* def name = get data //magician[@id='2']/name
* print "Magician with id 2 is", name
```

We get normal JSON data back when we use XPath, so matches such as this are not a problem:

```
* match data //magician[name = 'Harry Houdini']/specialty ==
['escapology', 'card tricks']
```

This one searches for any `magician` element containing a `name` element with the `Harry Houdini` value. It then retrieves all `specialty` element values from it. Since this magician has two specialties, they are returned as an array with two elements: `escapology` and `card tricks`.

Let's look at a similar example in JSON.

Working with JSONPath

JSONPath allows us to not only access certain JSON elements but also perform more complex searches and filtering. A great way to experiment with this is to use an online JSONPath test site such as `https://jsonpath.com`. This makes it possible to try out JSONPath expressions on your data first until they work correctly before integrating them into a Karate scenario.

JSONPath details

JSONPath expressions are comparable to XPath in that they refer to the JSON structure. Other than XML, there is mostly no root node that everything else is under but an *outer level*. This outer object is represented by a dollar symbol ($). However, you need to be careful when using the $ symbol in Karate, since here it represents the last API response. This was deliberately implemented so that you can directly execute JSONPath on Karate responses. So, depending on your context, this might not work as expected.

A nice full tutorial about JSONPath can be found here:

`https://goessner.net/articles/JsonPath/index.html`

Let's start with a JSON structure that is like the XML data we used before:

```
Scenario: JSONPath
  * def data =
  """
  {
      "magicians": [
        {
            "id":1,
            "name":"Harry Houdini",
            "birthyear":1874,
            "specialty":[
              "escapology",
              "card tricks"
            ]
        },
        {
```

```
            "id":2,
            "name":"David Copperfield",
            "birthyear":1956,
            "specialty":[
              "illusions"
            ]
          }
        ]
      }
    """
```

We can now implement similar checks as with XPath using JSONPath:

```
* match data.magicians == '#[2]'
```

We can use this shortcut to expect an array with two elements without using dedicated Karate functionality.

If we want to get the name of the magician where `id` is of value 2, this is one of the ways to go:

```
* def name = get[0] data.magicians[?(@.id==2)].name
* match name == 'David Copperfield'
```

There are some things to note:

- The `[?(@.id==2)` expression is a JSONPath filter statement with these parts:

    - `?()` is the syntax that encloses a filter function
    - `@` represents each array element, so this filter is applied to all `magicians` entries
    - `.id==2` is the actual filter we want to apply to each element

- After the filtering, we retrieve the `name` value.

- The result comes back as an array because there could potentially be many matching elements. In our case, we expect one, so to save only the first element in the Karate `name` variable, we need to use the `get` keyword again, but this time with an array index of 0.

As we did before, we can also do the same thing for the `specialty` keys:

```
* def specialties = get[0] data.magicians[?(@.name == 'Harry
Houdini')].specialty
* match specialties == ['escapology', 'card tricks']
```

This syntax might look complicated, but it also lets you express complex filtering in a very compact way. It's always a trade-off between readability and functionality, and it always comes down to the stakeholders and maintainers who need to work with and understand these tests.

Using karate.filter as an alternative

One of Karate's strengths is that there are always alternative approaches available. So, in the case of filtering an array by specific values, this can also be done with the built-in `karate.filter` function.

Taking the example from before, this is another way to filter for magicians whose specialties include *illusions*:

```
* def illusionistFilter = function(x){ return
    x.specialty.contains("illusions"); }
* def illusionists = karate.filter(data.magicians,
    illusionistFilter)
* print illusionists
```

Here, we define a function called `illusionistFilter` that can be used to filter by elements whose specialty array includes the `illusions` string. Inside the function, we are free to name the parameter as we like; here, I just called it x. Using `karate.filter`, we can then specify the dataset we want to filter (here, `data.magicians`) and our `illusionistFilter` function.

This gives us the correct result for our query:

```
09:44:32.904 [pool-1-thread-1] INFO  com.intuit.karate -
[print] [
  {
    "id": 1,
    "name": "Harry Houdini",
    "birthyear": 1874,
    "specialty": [
      "escapology",
      "card tricks"
    ]
  }
]
```

Here, you must decide between more code and more readability and then choose the appropriate approach.

> **karate.repeat**
>
> As a sidenote, the `karate.filter` mechanism of using a pre-defined function as an argument can also be used for data generation. The `karate.repeat` command takes a count and a function as arguments and can be used like a loop. It will execute the function as many times as specified by count and collect the results in an array.

Let's leave this topic aside for now and move on to the last topic of this chapter: how to test GraphQL APIs.

Testing GraphQL

GraphQL is a query language for APIs that was developed by Facebook. It provides a more efficient and flexible alternative to traditional REST APIs, which often require multiple requests to retrieve all the data that a client needs. With GraphQL, clients can specify exactly what data they need in a single request, and the server will return the requested data in a predictable and organized way. This allows for more efficient data transfer and reduces the amount of unnecessary data that is sent over the network. GraphQL is often used in modern web and mobile applications to improve performance and make it easier for developers to work with APIs.

> **More information**
>
> It is important to note that there is a lot more that GraphQL can do than what is mentioned here since the main scope of this book is to provide a thorough overview of Karate testing. If you want to know more about GraphQL, please see the official documentation here: `https://graphql.org`.

In this section, we will look at how we can test GraphQL endpoints using Karate. This concept can also potentially be carried over to other API formats that might pop up in the future and that Karate does not support natively (yet).

Understanding GraphQL requests

A GraphQL request is in a special, JSON-derived format sent via a *POST* request. This is not valid JSON, though, even if it looks similar at first glance. This makes its usage with Karate a little bit different than typical JSON-based REST APIs.

It always includes a `query` object in its request body that defines exactly which parts of the data should be returned. Also, it is possible to specify parameters in parentheses for each requested element, which would be invalid in pure JSON.

For this example, we will use **GraphQLZero** (`https://graphqlzero.almansi.me`), a public fake API that behaves like a real GraphQL API. It is a project that was inspired by the *JSONPlaceholder* API we used before in *Chapter 4* and uses a similar set of data: users, posts, and albums.

Take this query, for example. It returns the post that has an `id` value of `1` when we send it to the `https://graphqlzero.almansi.me/api` GraphQL endpoint:

```
query {
  post(id: 1) {
    id
    title
    body
  }
}
```

The nice thing now is that we can exactly specify what data we need. So, if `id`, `title`, and `body` are specified, this is exactly what we get—not more and not less. The return value is a regular JSON object containing this data. This is not contained in the root of the JSON but always with the `data` key:

```
{
    "data": {
        "post": {
            "id": "1",
            "title": "sunt aut facere repellat provident
occaecati excepturi optio
reprehenderit",
            "body": "quia et suscipit\nsuscipit recusandae
consequuntur expedita et cum\nreprehenderit molestiae ut quas
totam\nnostrum rerum est autem sunt rem eveniet architecto"
        }
    }
}
```

Note that only one single GraphQL request is needed even if we had nested data that would typically come from different requests to multiple endpoints. That means that this is a very efficient technology in terms of the amount of data and the number of requests that must be made.

Exploring the mock API

To check out which requests and responses are possible for the GraphQLZero API, it offers an online playground that lets you explore it and try out different requests. This can be accessed under the same URL that we use as our endpoint (`https://graphqlzero.almansi.me/api`):

Figure 6.6 – GraphQLZero playground

Here, you can write your requests on the left side and directly execute them with the play button. It also has complete documentation about all the different query possibilities and responses.

Also, you can use Postman to test this out. When sending a request body, it lets you specify GraphQL as a format, as seen in *Figure 6.7*. Then, it even highlights syntax errors so that you can spot these before sending a malformed request to the server. For GraphQL in particular, it is always good to thoroughly test out the requests before using them within your Karate tests because they can be hard to debug:

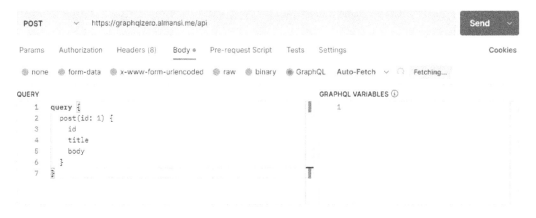

Figure 6.7 – Postman with GraphQL

Let's use this now in Karate.

Using GraphQL in Karate

For this example, we want to get all titles of a specific user's posts along with the name of this user. This could potentially be a valid request that is used for an overview web page of a user's posts:

```
Scenario: Query for user posts
  * url "https://graphqlzero.almansi.me/api"
  * text userPosts =
  """
  {
    user(id: 1) {
      name,
      posts {
        data {
          title
        }
      }
    }
  }
  """
  And request { query: '#(userPosts)'}
  When method post
  * print response.data.user.name
```

There are a couple of interesting things to note here:

- The GraphQL query to extract the username and post titles of user 1 is written as a Karate multiline expression, which makes it very readable.

- Since it does not follow the JSON standard, we must specify it with the text keyword so that Karate accepts it.

- The request { query: '#(userPosts)'} line where the query is constructed uses a Karate feature that makes it possible to use variables when constructing a JSON object. Here, the userPosts variable containing the query string is injected into our request body.

The response contains valid JSON inside the data element, so there is no need to do any further conversion:

```
{"data":{"user":{"name":"Leanne
Graham","posts":{"data":[{"title":"sunt aut facere
repellat provident occaecati excepturi optio
reprehenderit"},{"title":"qui est esse"},{"title":"ea
molestias quasi exercitationem repellat qui ipsa sit
aut"},{"title":"eum et est occaecati"},{"title":"nesciunt
quas odio"},{"title":"dolorem eum magni eos aperiam
quia"},{"title":"magnam facilis autem"},{"title":"dolorem
dolore est ipsam"},{"title":"nesciunt iure omnis dolorem
```

```
tempora et accusantium"},{"title":"optio molestias id quia
eum"}]}}}}
```

So, the `print response.data.user.name` line will print out Leanne Graham in this case.

Collecting data into a new array

In this case, since we are interested only in the titles of posts, it can make sense to flatten this structure into a simple array of strings. There is a very convenient way to do that, which again uses the power of custom functions in conjunction with built-in Karate object functionality.

Let's add a few lines to the bottom of our scenario:

1. This adds a new empty array called `titles` that should contain all post titles at the end:

    ```
    * def titles = []
    ```

2. Next is a variable called `getTitle` that defines a function that takes a `postData` parameter, extracts the `title` property from it, and adds it to the `titles` array from before. For this, we can use the `karate.appendTo.` built-in function:

    ```
    * def getTitle = function(postData) { karate.
    appendTo(titles, postData.title) }
    ```

3. Now, we use the `karate.forEach` function to iterate through all entries in the `response.data.user.posts.data` array and apply our `getTitle` method to each one of them:

    ```
    * karate.forEach(response.data.user.posts.data, getTitle)
    ```

If we now add a final `* print titles` line and run the scenario, this should achieve exactly what we want and lead to an array containing all post titles:

```
22:54:23.559 [pool-1-thread-1] INFO  com.intuit.karate -
[print] [
  "sunt aut facere repellat provident occaecati excepturi optio
reprehenderit",
  "qui est esse",
  "ea molestias quasi exercitationem repellat qui ipsa sit
aut",
  ...
]
```

You hopefully see that Karate is very future-proof in that it is well equipped to handle new approaches to APIs such as GraphQL with the `text` keyword and techniques such as variable substitutions. Also, special requirements—in this case, extracting the post titles and putting them into an array—can be solved easily by built-in Karate functionality in combination with small custom functions.

Summary

In this chapter, we have seen how to work with headers and cookies. We then looked at different new configuration and test run options that give us more flexibility and help to organize our test code better. An important Karate feature, defining expressions with the `def` keyword was the next topic in line, followed by JSONPath and XPath for efficient filtering of JSON and XML. Finally, we explored the GraphQL-specific approaches as an example of testing another data format apart from pure JSON.

In *Chapter 7, Customizing and Optimizing Karate Tests*, we will see how Karate can be extended with more custom functionality. Also, we will take a closer look at how we can make our code even more concise and easier to maintain.

7

Customizing and Optimizing Karate Tests

Karate already offers so many built-in features and functionality that it is impossible to cover everything in one book. However, sometimes it can be necessary to add more custom functionality to it. This typically happens if there are different data sources involved in test cases that go beyond APIs, or if test cases deal with data formats that Karate does not support natively.

Also, depending on your use case, mocking certain data might be required to not depend on third-party provider availability. Luckily, there are mechanisms to deal with all of this in a straightforward way.

Until now, we have written mostly shorter test scenarios that are not too hard to maintain in the long run. Typically, test suites can get very big the more test cases are added and coverage is achieved. Therefore, it is necessary to think about how to keep tests short, concise, and maintainable. Plus, this has the added benefit of increased understandability.

In this chapter, we will look at both aspects by covering these main topics:

- Using Karate hooks
- Defining and calling Java functions
- Using Karate as a mock server
- Making your tests more concise

Technical requirements

The code examples for this chapter can be found at https://github.com/PacktPublishing/Writing-API-Tests-with-Karate/tree/main/chapter07.

You will require the following:

- The system and IDE setup we completed in *Chapter 2, Setting Up Your Karate Project*
- Postman to explore our mock example
- A MySQL database (if you do not have a local one, I will explain how to set up a free online instance)

We will first look at the topic of Karate hooks, which offer a simple way to react to specific events.

Using Karate hooks

Hooks are a special mechanism to allow events to be received from test runs and run code in reaction to them. In this example, we will use hooks to output some information about scenarios and steps while tests are executed.

To use hooks, you need to do two things:

1. Implement a `Java` class implementing Karate's hook interface: `com.intuit.karate.RuntimeHook`.
2. Register the new hook class as a Karate hook in your runner class.

In the following sections, we go through both steps.

Implementing a new hook class

In our example, we want to add some more log outputs that tell us which scenario is started and finished and what each step result is. For this, we can start with a class implementing Karate's `RuntimeHook` interface like this:

```
package hooks;

import com.intuit.karate.RuntimeHook;

public class KarateHooks implements RuntimeHook {
}
```

In our example project, this class is called `KarateHooks` and resides in the `hooks` package on the same level as our usual `examples` package, so directly under `src/test/java`.

Figure 7.1 – Project setup

This is already a valid hook class, even though it does not do anything yet. So, the next step is overriding methods that are triggered when we want to react to certain Karate test events.

In *Figure 7.2*, you can see all the available methods in Karate 1.2.0 that I am using here (in newer ones, there might be even more). If you use VS Code, you can press *Ctrl* + the spacebar to display this list and select a method to override (or right-click and select **Source Action… | Override/Implement Methods…**).

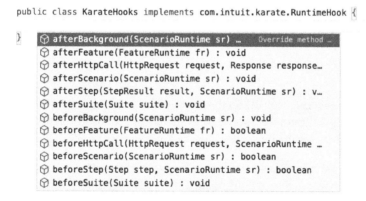

Figure 7.2 – Available hook methods

Most of these methods are test scenario-related. There is one exception: `beforeHttpCall()` and `afterHttpCall()` can be used to react to web service calls or even intercept and change them before they are performed.

In our example, we will use some of the other methods, though, as they show the overall concept more clearly.

Overriding handler methods

Let's first react to a starting scenario to get a feel of the general mechanism of these methods. From the list of methods, we can pick the conveniently named `beforeScenario` method for this:

```
@Override
public boolean beforeScenario(ScenarioRuntime sr) {
    System.out.println("Scenario is starting: "
        + sr.scenario.getName());
    return RuntimeHook.super.beforeScenario(sr);
}
```

There are a few points to note here:

- The overridden `beforeScenario` method accepts an argument of the `ScenarioRuntime` type. This holds all the information about the scenario itself, its caller, tags, the feature it is included in, and lots more. It is worthwhile to examine the Karate source code for such cases. Here, you can find many helpful methods and properties that you can use.

- The overridden method should call the original hook in the end (in this case, `RuntimeHook.super.beforeScenario(sr)`). Otherwise, you might run into strange behaviors while tests are running. When you use VS Code's feature to override methods, this is automatically generated for you.

- This new method prints out `Scenario is starting:` and the name of the current scenario taken from the passed `ScenarioRuntime` instance by calling `sr.scenario.getName()`.

- You might be wondering why the method returns a `boolean` value. This is because, in the case of `false`, it indicates that this scenario should be excluded from the test run. We don't have to worry about setting this as it is done by Karate automatically.

We can add the rest of the methods we need along the same lines. Let's add one like the one before for `afterScenario`, which should notify us that the scenario has finished running and provide the result:

```
@Override
public void afterScenario(ScenarioRuntime sr) {
    System.out.println("Scenario finished: "
        + sr.scenario.getName()
        + ", result: " + sr.result);
    RuntimeHook.super.afterScenario(sr);
}
```

One difference here is that this method does not return a value because it is only executed in the case that the scenario runs at all. Also, we can print `sr.result` here, which gives us the result of the whole scenario.

Let us add one more method that is executed for each step as well:

```
@Override
public void afterStep(StepResult result, ScenarioRuntime sr) {
    if (result.isFailed()) {
        System.out.println("Step failed: "
            + result.getStep().getText());
    } else {
        System.out.println("Step passed: "
            + result.getStep().getText());
    }
    RuntimeHook.super.afterStep(result, sr);
}
```

The signature of the `afterStep` method is different in that we have both `ScenarioRuntime` for the whole scenario and `StepResult`, which includes some more step-related information. We use the `isFailed()` method here to determine whether the current step passed or failed. Note that, again, this method calls the original `RuntimeHook.super.afterStep(result, sr)` method in the end.

Registering a hook class in the runner

The next part is to let the Karate runner know that we want to use a hook class as this is not done automatically. This even makes it possible to have multiple hook classes or switch out the hooks you want to use in a certain test run.

For registration, you can use the `hook()` method. This is available in the `@Karate.Test` annotated run methods:

```
@Karate.Test
Karate testUsers() {
    return Karate.run("hooks")
        .hook(new KarateHooks()).relativeTo(getClass());
}
```

If you use a parallel runner, the hook() method is there as well:

```
@Test
void testParallel() {
    Results results = Runner.path("classpath:examples")
        .hook(new KarateHooks()).parallel(5);
    assertEquals(0, results.getFailCount(),
        results.getErrorMessages());
}
```

Note that we do not register the KarateHooks class itself but a new *instance* of it.

Running a test using the hook class

Let's take a very simple test scenario that only prints out the first two step names and does a failing match afterward:

```
Scenario: Hooks demo
    * print "This is step 1"
    * print "This is step 2"
    * match 1 == 2
```

If we run this scenario now with the registered hook, we get this output:

```
Scenario is starting: Hooks demo
22:17:13.346 [pool-1-thread-1] INFO  com.intuit.karate -
[print] This is step 1
Step passed: print "This is step 1"
22:17:13.358 [pool-1-thread-1] INFO  com.intuit.karate -
[print] This is step 2
Step passed: print "This is step 2"
22:17:13.386 [pool-1-thread-1] ERROR com.intuit.karate -src/
test/java/examples/hooks/hooks.feature:6
* match 1 == 2
match failed: EQUALS
  $ | not equal (NUMBER:NUMBER)
  1
  2
```

```
src/test/java/examples/hooks/hooks.feature:6
Step failed: match 1 == 2
Scenario finished: Hooks demo, result: [failed] * match 1 == 2
```

You can see that both the scenario-related *before* and *after* hooks as well as the *after-step* hook work correctly and give us the expected output.

Defining hooks inside feature files

There is another way to define hooks when they should be scenario- or feature-specific. These can be included inside the feature files and do not require any Java code:

```
Background:
    * configure afterScenario = function(){ karate.log(karate.
scenario.name + ' finished!') }
    * configure afterFeature = function(){ karate.log(karate.
feature.fileName + ' finished!')
    }
```

By configuring specific `afterScenario` or `afterFeature` functions in the background, these are automatically called by Karate when a scenario or feature finishes, respectively. In this case, we log the name of the scenario when a scenario ends and the filename of the feature file when it finishes running using the `karate.log` function.

When we run our scenario now, we should see this new output at the very end of the logs:

```
13:57:58.117 [pool-1-thread-1] INFO  com.intuit.karate - Hooks
demo finished!
Scenario finished: Hooks demo, result: [failed] * match 1 == 2
13:57:58.119 [pool-2-thread-1] INFO  com.intuit.karate - hooks.
feature finished!
```

Of course, you could also configure these inside of `karate-config.js` so they are available across all features. Note that these hooks are only called when a scenario is executed as a test, not in scenarios called from other scenarios – for example, as data providers or setup code.

These two mentioned hook approaches can be very useful if you need to react to specific events in all your tests. However, if you want to extend Karate even further by functionality, there is a very powerful way that we will explore next.

Defining and calling Java functions

Even though JavaScript offers a lot of functionality, sometimes it can be easier to use Java for specific tasks. Fortunately, Karate can call and use Java classes and methods within JavaScript, as mentioned in *Chapter 1, Karate's Core Concepts.*

Understanding the basics

First, let's see how to use Java classes that are part of Java itself to slowly start exploring this powerful feature. For this example, I want to determine the current directory of our running test. There are, of course, many ways to do this, both in JavaScript and Java. Here, I will use the `java.nio.file.Paths` class that is built into Java to accomplish this.

This would be the pure Java statement:

```
String currentPath =
    Paths.get(".").toAbsolutePath().toString();
```

Basically, this gets a `Path` instance of `"."` (which stands for the current path) and then returns the full path of this using the `toAbsolutePath()` method. Since we want to have the result as a string, we need to call `toString()` here as well.

Using the same functionality within a Karate scenario is rather straightforward with Karate's `Java.type` method:

```
Scenario: Simple Java access
    * def Paths = Java.type('java.nio.file.Paths')
    * def currentPath = Paths.get(".").toAbsolutePath()
    * print currentPath
```

This is how it works step by step:

1. We use `Java.type('java.nio.file.Paths')` to get access to the Java class we want to use. It is important to use the fully qualified class name including its path.

2. This reference is stored in the JavaScript `Paths` variable. This acts like a bridge to the Java class.

3. Now, we can use the JavaScript `Paths` variable to call the same methods we would in a pure Java program (`get(".").toAbsolutePath()`).

4. The result can then be assigned to a new JavaScript variable, `currentPath`. We don't need the `toString()` call here as JavaScript is less strict than Java.

If everything is correct, running this scenario will output the path of wherever you are running it from. In my case, it is as follows:

```
13:03:26.280 [pool-1-thread-1] INFO  com.intuit.karate -
[print]
/Users/bbischoff/Development/Writing-API-Tests-with-Karate/
chapter07/java/.
```

Let's now see how we can use more advanced custom Java functionality by the same means as before.

Working with databases

For this example, we want to retrieve a list of famous magicians from a database so we can use this data source in our tests. This can be a real-world scenario in which APIs and databases are used in combination. Please note that this example only deals with reading from a database table. It can also be extended to support writing to a database should you desire to add this later.

Setting up the database

In this example, we will use Java to connect and read from a MySQL database. You can either use a local MySQL installation or a free online one. In this section, I will guide you through its setup.

We will use the `https://freesqldatabase.com` service where you can create MySQL test instances free of charge. Follow these steps:

1. Scroll down on the home page and click on the **Sign Up** button for the **Free** tier:

Figure 7.3 – FreeSQLdatabase sign up

2. After registering, you need to check your email for verification. The actual database creation process starts with your choice of a server location near you. Since I am in Germany, I chose Europe here.

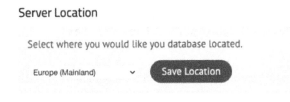

Figure 7.4 – FreeSQLdatabase database server location

3. After clicking the **Start new database** button in the next step, your new database will be prepared.

Figure 7.5 – FreeSQLdatabase database spin up

When it is done, you will receive a new email with the host, DB name, and DB password.

4. Now you can go to `https://github.com/PacktPublishing/Writing-API-Tests-with-Karate/blob/main/chapter07/java/src/test/resources/magicians.sql` and save the contents to a new file on your computer. These are the SQL statements that create the database table we need for our example.

5. Head to `https://www.phpmyadmin.co`. Here you can find a complete online phpMyAdmin instance that allows you to manage your MySQL database. After entering the credentials that you received earlier from **FreeSQLdatabase**, you will be presented with the **phpMyAdmin** user interface.

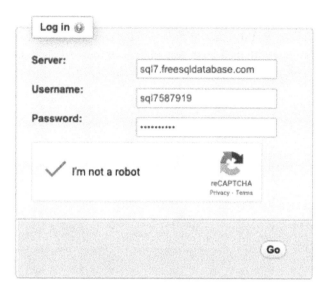

Figure 7.6 – phpMyAdmin log in

Please pay attention to entering the host (in **phpMyAdmin**, it is called **Server**) without a protocol such as `http://` or `https://`!

6. The only thing we need to do here is create our database table that we want to fill with the prepared SQL statements. Proceed to the **Import** tab and click on **Choose file** to select your local SQL file from before.

Figure 7.7 – phpMyAdmin file import

7. Click the **Go** button to start the import. After this, you should see a new `magicians` table
    in the sidebar. Clicking on this one will show you the contents that are stored there – some
    famous magicians along with their birth year and one or more specialties.

Figure 7.8 – Created "magicians" table in phpMyAdmin

This is all we need to do in phpMyAdmin. The rest will be done in the IDE.

Setting up the project

We will start with a new project based on the Karate archetype that we used in all the former chapters.
To be able to communicate with a database through Java, we need to add an additional library.

Let's add the `mysql-connector-j` dependency to our `pom.xml` file's `<dependencies>` block
so that Maven can download and add this to our project automatically. This enables communication
with a MySQL database through Java (more information can be found at `https://github.com/`
`mysql/mysql-connector-j`):

```
<dependency>
    <groupId>com.mysql</groupId>
    <artifactId>mysql-connector-j</artifactId>
    <version>8.0.31</version>
</dependency>
```

Next up is writing the code to access and read from our demo database.

Writing a database class

Now that we have the dependency in place, we can start writing a class that can access our
`magicians` database.

Figure 7.9 shows our final project setup, so the Java classes needed for the DB access and dealing with the mapping to JSON will be in a db package below `src/test/java`:

Figure 7.9 – Database project setup

Let's start with the MySQL.java class in the db package, which will be our bridge to the database. This is the complete code of this class:

```java
package db;
import java.sql.Connection;
import java.sql.DriverManager;
import java.sql.ResultSet;
import java.sql.Statement;
import java.util.ArrayList;
import java.util.List;
import com.intuit.karate.JsonUtils;

public class MySQL {
    public String getMagicians() throws Exception {
        String connectionString =
            "jdbc:mysql://sql7.freesqldatabase.com/sqlxxx";
        String user = "sqlxxx";
        String password = "xxx";
        try (
            Connection c = DriverManager.getConnection(
                connectionString, user, password
```

```
        );
        Statement s = connect.createStatement();
        ResultSet result = statement.executeQuery(
            "select * from magicians"
        )
    ) {
        List<Magician> magicians = new ArrayList<>();
        while (result.next()) {
            String name = result.getString("name");
            int birthyear = result.getInt("birthyear");
            magicians.add(
                new Magician(name, birthyear)
            );
        }
        return JsonUtils.toJson(magicians);
    }
}
}
```

It might look overwhelming at first, but we will break this down into its individual parts and examine them one by one. Remember that this is the only code we need to access the database and get the data we want.

> **A word about the demo code**
>
> Please note that this demo class is not a fully worked-out solution for dealing with databases. To use it in a real project, it would need some more code to make it more secure, flexible, and stable. However, that is not the goal here.

Let's start with a getMagicians() method now. This should return a string in the end that includes all magicians that are read from the database. We will make sure this string contains valid JSON so that we can work with this efficiently in Karate.

First, we define a new **Java Database Connectivity (JDBC)** string. This is a standard format for Java to connect to a database:

```
String connectionString =
    "jdbc:mysql://sql7.freesqldatabase.com/sqlxxx";
```

Let's look at the individual parts:

1. The string starts with `jdbc:` indicating that this is a JDBC connection string.

2. `mysql://` specifies the database type we want to connect to.

3. `sql7.freesqldatabase.com` is the database host. Note that you indicate this without a preceding protocol. If you are using a local database, this is usually `localhost` or `127.0.0.1`.

4. Finally, `/sqlxxx` points to the actual database (you need to replace `sqlxxx` with your assigned database name).

Even though we could also add the username and password to this string as parameters, it is clearer to keep them in separate variables. Here again, you need to enter your given database credentials:

```
String user = "sqlxxx";
String password = "xxx";
```

It is usually not good practice to keep credentials like this in the source code but for this demo, it is okay. We will see how we can handle this better in *Chapter 8, Karate in Docker and CI/CD Pipelines*.

Now, we can use this in a so-called `try` resource block. This means that everything within the following block will only be executed when the resources specified in the parentheses of the `try` block can be created successfully. Also, this approach automatically closes all opened connections after the following code block is finished, which avoids memory leaks and other problems:

```
try (
    Connection c = DriverManager.getConnection(
        connectionString, user, password
    );
    Statement s = c.createStatement();
    ResultSet result = s.executeQuery(
        "select * from magicians"
    )
) {
    // This code here will be explained in the next section
}
```

First, we establish a new connection to our database using `connectionString, user`, and `password`. This is done using the `DriverManager.getConnection` method, which accepts these three parameters. We save this in the `c` variable.

Next is a `Statement` instance called `s` that we can create from our `c` connection with `c.createStatement()`. This can be used to next execute a **Structured Query Language (SQL)** statement against our database.

The statement we will use is `select * from magicians`, meaning "retrieve everything from the `magicians` table." Executing this with the `executeQuery` method from our `Statement` instance, `s`, ensures that it is sent to the correct database.

> SQL
>
> If you want to know more about SQL, specifically the SQL supported by MySQL, a good resource is the official MySQL reference at `https://dev.mysql.com/doc/refman/8.0/en/sql-statements.html`.

The `executeQuery` method returns a `ResultSet` instance, which we will call `result`. This includes all database rows that are returned by our `select` statement. Next, we will turn this into a format that we can work with more easily within Karate.

Adding a POJO

The acronym **POJO** stands for **Plain Old Java Object** and is just a fancy name for a class holding nothing but data. We will create a new class called `Magician` in the existing `db` package. We can use this later to facilitate turning `ResultSet` into JSON:

```java
package db;

public class Magician {
    private String name;
    private int birthyear;

    public Magician(String name, int birthyear) {
        this.name = name;
        this.birthyear = birthyear;
    }

    public String getName() {
        return name;
    }
    public int getBirthyear() {
        return birthyear;
```

```
    }
}
```

In a nutshell, this class can hold two values: name and birthyear. We will ignore everything else for now to keep this demo as simple as possible.

A key point here is that we can create a new Magician instance that holds our passed values. For example, this statement would create a new magician with the name Peter and a birth year of 1980:

```
Magician magician = new Magician("Peter", 1980);
```

Because name and birthyear are declared as private, we need to add a so-called getter method for each so we can access them from the outside. These methods, getName() and getBirthyear(), do nothing more than return these respective values.

Turning ResultSet into JSON

Now, let's get back to the inside of our try block that created Connection, Statement, and ResultSet:

```
List<Magician> magicians = new ArrayList<>();
while (result.next()) {
    String name = result.getString("name");
    int birthyear = result.getInt("birthyear");
    magicians.add(
        new Magician(name, birthyear)
    );
}
return JsonUtils.toJson(magicians);
```

Here, we do the following:

1. Create a new empty list that can hold Magician instances:

    ```
    List<Magician> magicians = new ArrayList<>();
    ```

2. Then, we go through each row in the returned ResultSet using its result.next() method. This is the only way to loop through it:

    ```
    while (result.next()) {
    }
    ```

3. Inside this loop, we retrieve and store the two values we need (name and birthyear) by referencing the respective database table columns by their names:

```
String name = result.getString("name");
int birthyear = result.getInt("birthyear");
```

Note that we need to use the correct ResultSet method based on the data type of the MySQL column: getString for a string value and getInt for a numeric value.

4. Now we can create a new Magician instance and add it to our list of magicians:

```
magicians.add(
    new Magician(name, birthyear)
);
```

5. Karate works best with JSON values, so after filling our list of magicians, we can use the toJson() method from Karate's JsonUtils class to turn this list into JSON!

Using the database functionality in a test scenario

Now, we can use everything in a Karate scenario:

```
Scenario: MySQL access
  * def getMagicians =
    """
    function() {
      var MySQL = Java.type('db.MySQL');
      var mySQL = new MySQL();
      return mySQL.getMagicians();
    }
    """
  * json magicians = getMagicians()
  * print magicians[0]
```

As with the simple Java example from the beginning, we start with defining a variable, getMagicians. This points to a function that does the following:

1. Defines a variable called MySQL (notice the capital *M*) that points to our new db.MySQL class using Karate's Java.type function

2. Creates an instance of this class and stores it in a variable called mySQL (with a lowercase *m* this time)

3. Returns the JSON result of the getMagicians() method of our Java class

At this point, nothing is executed yet. So far, we only prepared everything for later execution.

The `* json magicians = getMagicians()` step executes the function, calls our Java class, returns the JSON result, and stores it in the `magicians` variable. It is important to note that the method returns JSON format as a string, so we need to use Karate's `json` method to turn this into a real JSON object.

Running this scenario, we get the expected value of the first magician in the database:

```
16:49:39.307 [pool-1-thread-1] INFO  com.intuit.karate -
[print] {
  "birthyear": 1874,
  "name": "Harry Houdini"
}
```

Hopefully, this example has shown how powerful custom Java code can be. It opens tons of additional functionality that can go far beyond what Karate natively supports. Fortunately, it's not that difficult to combine JavaScript and Java in the Karate context.

Next, we will look at a completely different use case of Karate: mocking APIs!

Using Karate as a mock server

API mocks (or test doubles, as Karate prefers to call them) are pieces of code that simulate APIs that are dependencies of certain test scenarios. By using these instead of the real APIs, tests can run much faster and more stable since these mocks are always available and provide exactly the data that is expected. Also, if tests need to be written even though certain APIs are only specified but not implemented yet, mocks are a great choice. Later, they can potentially be replaced by real implementations.

A great point is that these mocks can keep their state. This means it is potentially possible to simulate full **Create, Read, Update, Delete** (**CRUD**) APIs that can add, get, edit, and remove data.

Authoring a mock scenario

Karate has a very straightforward built-in way to define and use mocks by employing nothing more than the same conventions that are used to write test scenarios. Let's see what this means.

For our example, we will mock an API that retrieves magicians as JSON objects. For this, we start with a regular feature file:

```
Feature: Mock server

  Background:
    * def magicians =
```

```
    """
    [
        { id: 1, name: "Penn" },
        { id: 2, name: "Teller" }
    ]
    """
    * json magicians = magicians
```

We start with a Background in which we define a JSON structure of magicians. Each object in our magicians array has two values, id and name, linked to it. Since we use the multi-line expression syntax to define our JSON that returns the structure as a string, we need to use Karate's json command to turn this back into a real JSON object.

This feature file will include the definitions for our mocked API. Each scenario included here will not represent a test but a specific API endpoint in this case.

Defining the /magicians endpoint

Let's start with the /magicians endpoint. When doing a GET request, we want it to return the complete magicians JSON object as a response:

```
Scenario: pathMatches('/magicians') && methodIs('get')
    * def response = magicians
```

This is the complete scenario for this endpoint. You can see that this scenario does not have a name. Instead, there is logic in its place. pathMatches('/magicians') means that this scenario is executed when the /magicians endpoint is used in a request. The methodIs('get') condition specifies that this should only happen on GET requests. If this is omitted, this scenario would be triggered on all types of requests, for example, POST, PUT, and so on.

The response variable is very special here. Whatever we set it to will be what this API request returns. Here, we set it to the magicians object that we previously defined in the Background scenario.

Let's add a more complex one that contains more logic inside.

Defining the /magician endpoint

This scenario is supposed to react to GET requests such as /magician/0 or /magician/1 to select one of our magicians by their ID. This is done with pathMatches('/magician/{id}'):

```
Scenario: pathMatches('/magician/{id}') && methodIs('get')
    * def index = parseInt(pathParams.id) - 1
    * def isValid = index > -1 && index < magicians.length
```

```
* def responseDelay = 3000
* def response = isValid ? magicians[index] : ''
* def responseStatus = isValid ? 200 : 204
```

Here, {id} acts as a path parameter that can have any string value. In the next line, we can retrieve this id parameter by using pathParams.id.

Since this is a mock, we don't have to develop a very sophisticated logic if this API delivers the results we want. Therefore, we can turn this id parameter into a number with parseInt and subtract *1* so we get the correct array index for our magicians array – for example, requesting id with a value of 2 would retrieve the magician at array index 1.

As a small safety measure, we define an isValid variable that is true when the index variable is a number greater than *-1* and less than the number of available magicians.

One of the nice features of using a Karate mock is that it is possible to set a responseDelay variable to simulate a slow connection to the API. In this case, we set it to 3000 milliseconds. That's it, no further actions are necessary!

Also, we set the response variable as before, but this time, based on our isValid variable. If we have a valid index, we set response to magicians[index] to pick the magician with the correct id. Otherwise, it is set to an empty string. Again, setting this special variable is the only thing we need to do to define what should be returned.

Finally, we also set the special responseStatus variable that defines which HTTP status code should be returned. If we don't assign it a status code, it is 200 (OK) by default. Here, we only want it to be 200 if we have a magician to return, otherwise, we return 204 (No Content) to indicate that the requested magician ID does not exist.

Catch-all scenario

As the last part of our mock, we can define a scenario that should be invoked on any other unhandled request:

```
Scenario:
  * def responseStatus = 500
```

This one just sets responseStatus of 500 (Internal Server Error) for test purposes.

Now we are finally ready to put our mock to use!

If we try to run this mock definition scenario as a normal Karate test, it will fail. Its only purpose is to act as a description of mocks and it must be started in a special way.

Firing up the mock server from within a test

To see whether our implementation is working, we can test it with a normal Karate test scenario:

```
Feature: Testing a mock server
    Background:
        * def start =
                () => karate.start('mocks.feature').port
        * def port = callonce start
        * url 'http://localhost:' + port
```

Here, in our new feature file, we have a `Background` scenario that starts the mock server so that it is available when the tests start. It uses a so-called **arrow function**, which is a shorthand way of writing a traditional function, and assigns this to the `start` variable.

The `karate.start` method inside of this function accepts the name of a feature file that includes the API mocks we want to be served, so we point it to our `mocks.feature` file. When called, it starts up a mock server. Since we cannot know on which port it will start as there can be multiple ones running at the same time, we need to return the port number of this server using `.port`.

Now we can use Karate's `callonce` method followed by the method we want to trigger, in our case, `start`. The `callonce` invocation makes sure that this only happens once so we won't have a new server starting up for each test scenario. Also, we define a new variable called `port` to store the current port number of our test server. We know that this server runs on our local system, so we can use `'http://localhost:' + port` as our base URL for all scenarios.

Let's test it with a scenario checking our `/magician/{id}` endpoint from before. We already set the base URL in the background, so all that's left here is calling it and matching the response, as we have done multiple times now:

```
Scenario: Get one magician
    When path 'magician/2'
    And method get
    * match response.name == 'Teller'
```

If we run this now, we will see the test server starting up:

```
14:41:39.087 [main] INFO  c.l.armeria.common.util.SystemInfo -
hostname: benjaminbis2bbe (from 'hostname' command)
14:41:39.458 [armeria-boss-http-*:62585] INFO  com.
linecorp.armeria.server.Server - Serving HTTP at /
[0:0:0:0:0:0:0:0]:62585 - http://127.0.0.1:62585/
```

```
14:41:39.464 [main] DEBUG com.intuit.karate.http.HttpServer -
server started: benjaminbis2bbe:62585
```

After that, the test scenario is executed, resulting in a successful run and the mock server is stopped when the end of the test run is reached. We can also verify this in the logs, as the call to /magician/2 returns {"id":2,"name":"Teller"} as expected.

We can also add a scenario that tests our catch-all mock scenario like this:

```
Scenario: Unavailable endpoint
    When path 'rabbits'
    And method get
    Then match responseStatus == 500
```

Giving it the rabbits path, which is, of course, not existing in our mock definitions, we get back the anticipated status code of 500!

Using a standalone mock server

Executing a mock server from inside of your test code can be beneficial in some use cases – for example, when a mock server should be used together with a real API. It is also possible to spin up a standalone mock server that is independent of your test scenarios and test runs. This way, it can act as a more permanent mock so it can be reused across multiple different test suites or even multiple applications under test.

The easiest way to do this is to use Karate standalone, as discussed in *Chapter 2, Running Karate Tests*. Using the -m flag, we can pass our mocks feature to it:

```
java -jar
    c:\Users\bbischoff\Desktop\karate-1.2.1.RC1\karate.jar
        -m src/test/java/examples/mockserver/mocks.feature
```

Invoking this, Karate standalone will start up the mock server using our mock definitions:

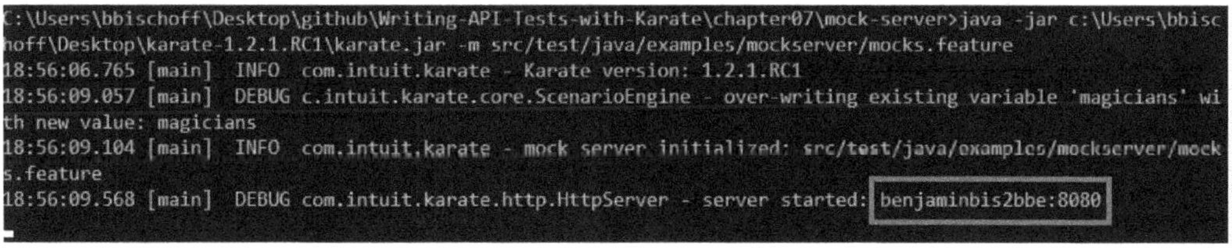

Figure 7.10 – Mock server startup with Karate standalone

Here, I started this server in a separate Windows Command Prompt window, so it is independent of VS Code. If you use a standalone server, you don't need the server startup code in your test feature files. Also, when starting it via Karate standalone, the mock server usually has port 8080, so the base URL to be used in the tests is `http://localhost:8080` and you don't have to deal with a dynamic port:

```
Feature: Testing a mock server
    Background:
        * url 'http://localhost:8080'
```

Note that the mock server is not stopped automatically like when it is started from inside the test scenario. To stop a standalone server, you can either press **CTRL-C** on the command line or close the terminal window completely.

Now that we have seen a way to use Karate for serving an API instead of testing it, let's get back to testing again. The following section will focus on how tests can more efficient and compact.

Making your tests more concise

There are certain principles in software development, such as **don't repeat yourself (DRY)** and **keep it simple, stupid (KISS)**, that lead to better and more maintainable code. If your code adheres to the *DRY* principle, it means that there is not a lot of duplicated code, whereas *KISS* means that your code should be easy to follow and understand. Both principles can be tackled in Karate, making your tests more fun to work with in the process.

Reducing code by calling other feature files

`Background` scenarios solve code duplication *inside a single feature* file, as we saw before. However, it can be beneficial to define common functionality *for multiple feature files* in a central place.

Also, it is possible to split feature files into smaller reusable parts so that we don't have very large scenarios that are hard to read and understand. In this section, we will cover both cases.

We will start with a feature file called `sayhello.feature` that only has one single scenario. In fact, this is the recommended approach if you create a callable feature:

```
@ignore @report=false
Feature: Say hello feature
    Scenario: Greet a person
        * print "Hello", name
```

For demonstration purposes, it will just output `Hello` and a `name` variable. Note that this variable is not defined here but will eventually come from the calling scenario. We don't want this feature or the enclosed scenario to be invoked on its own because it is a helper feature. Also, calling it directly would fail because of the missing variable definition. That's why we add two tags here. The `@ignore` tag prevents it from being invoked, whereas the `@report=false` tag hides its steps from the test report.

We can now use this feature from a test scenario that is in another feature file, like this one here:

```
Scenario: Access the name automatically
  * def name = "Benjamin"
  * call read('sayhello.feature')
```

Here, we use the `read` keyword to retrieve the contents of our helper feature. We could now store this in a variable to call it later. In this case, we call it directly by putting the `call` keyword in front of it. Note that before calling it, we define a `name` variable here. This is then automatically available within the called feature, so we get this output:

```
17:51:33.694 [pool-1-thread-1] INFO  com.intuit.karate -
[print] Hello Benjamin
```

It is also possible to pass a JSON parameter to the call of a helper feature. In this case, the `name` variable that is expected in the feature file needs to be included as a key:

```
Scenario: Pass name as param
  * def params = {name: "Anne"}
  * call read('sayhello.feature') params
```

Running this gives us the expected output, even though, this time, the `name` value is taken from the JSON:

```
17:55:04.962 [pool-1-thread-1] INFO  com.intuit.karate -
[print] Hello Anne
```

As a third alternative, you can also access the raw parameter within the called feature. This can be helpful if you deliberately want to pass JSON to it and work directly on this JSON inside the feature. To demonstrate it, I will add another scenario like the previous one that calls a different feature:

```
Scenario: Passing the JSON structure
  * def params = {name: "Frida"}
  * call read('sayhelloArg.feature') params
```

The `sayhelloArg.` feature file looks like this:

```
@ignore @report=false
Feature: Say hello with __arg
```

```
Scenario: Greet a person
    * print "Hello", __arg
```

The special `__arg` variable contains the exact parameter that was passed to this feature. So, in this instance, our output is different as it shows the passed-in JSON:

```
18:00:41.863 [pool-1-thread-1] INFO  com.intuit.karate -
[print] Hello {
  "name": "Frida"
}
```

In all these examples, we used the `call` keyword, meaning that each scenario freshly calls the helper feature file. If you don't want this, there is also the `callonce` keyword that executes only once and serves cached results on every subsequent invocation.

Let's see how we can further optimize our tests using data-driven approaches next.

Data-driven tests with scenario outline

If multiple scenarios have a set of similar steps, we can run into **code duplication**. The problem here is that if we decide to change something in these steps, we must do this for each scenario individually. This is sometimes called **shotgun surgery** since we apply the same small change multiple times in multiple places.

Data-driven testing is a method of software testing where test cases are derived from data. This approach could mean creating many test cases, with each test case containing a unique set of input data and the expected output. Luckily, Karate contains a lot of ways to make this process of data-driven test creation much quicker by reusing the same test code and providing an additional data source that is processed and turned into more elaborate test cases.

Avoiding code duplication with background scenarios

We can define common steps as a `Background` that is executed before any of the scenarios inside the same feature file. We had this before in some examples, when we moved common steps to it – for example, setting a base URL and making an initial request.

Here is a quick refresher on how this could look:

```
Scenario Outline: Normal outline
    * print '<name>'
    Examples:
        | name    |
```

```
| Eamon |
| Ginny |
```

In this scenario, we define an example table containing different values in each row (in case you wonder, these are names of pets I know). The first row contains the header of the column – here, name. For each of the rows, the scenario is run independently. In our case, we have two names, Eamon and Ginny, so Karate will turn this into two scenarios. We can print the name using the <name> notation referring to the appropriate column. So, our output, in this case, is as follows:

```
21:55:11.549 [pool-1-thread-2] INFO  com.intuit.karate -
[print] Ginny
21:55:11.549 [pool-1-thread-1] INFO  com.intuit.karate -
[print] Eamon
```

This is fine if you have a fixed set of test data, but Karate offers way more functionality here to make these scenario outlines more dynamic.

Using JSON in the example table

Usually, in Cucumber, example tables are limited to strings or primitive data types. If you want to change this, you will need to write custom **table transformer** classes that turn specific columns into other data types.

Karate, however, can directly work with JSON here:

```
Scenario Outline: JSON in data table
    * def color = traits.color
    * print '<name> is', color
    Examples:
        | name  | traits!          |
        | Eamon | {color: 'red'}   |
        | Ginny | {color: 'black'} |
```

In this case, our example table has two columns, name as before, and traits which contains JSON structures. These contain color keys that point to a color value of each animal. In larger tests, these structures could be much more complex.

If we want to use the traits column values as JSON, we must add an exclamation mark to the name traits!. This signals Karate that we want it to turn the values in this column into the appropriate data type. Otherwise, it would be considered as a string, and accessing traits.color would give us an error.

Running this scenario gives us the same kind of output as before, but now including the colors:

```
22:17:41.548 [pool-1-thread-1] INFO  com.intuit.karate -
[print] Eamon is red
22:17:41.548 [pool-1-thread-2] INFO  com.intuit.karate -
[print] Ginny is black
```

Now we can go a step further!

Using external files as data sources

Example tables can get quite big if you have large sets of data. A better way in this case is the use of data files. Since we saw that we can use JSON here, why not combine this with Karate's ability to read from files?

Let's use a file called `animals.json` that has the following JSON structure inside:

```
[
    {"name": "Ginny", "color": "black"},
    {"name": "Eamon", "color": "red"}
]
```

This is basically all the data from the example table in the last scenario, but now in a single structure. The good thing about a file like this is that we can use it right within an example table:

```
Scenario Outline: Data from file
    * print name, 'is', color
    Examples:
          | read('animals.json') |
```

Just by using the `read` function with the filename in a single example table cell, Karate will turn this JSON structure into variables that are directly usable within the test scenario. That means we can directly access `name` and `color` as these are the keys in the JSON file.

Note that this is still executed as two separate scenarios, giving us this output:

```
22:04:23.388 [pool-1-thread-1] INFO  com.intuit.karate -
[print] Eamon is red
22:04:23.389 [pool-1-thread-2] INFO  com.intuit.karate -
[print] Ginny is black
```

This is great, but we can do even better!

Generating test data from a feature

Yet another way of specifying test data is the use of specific features that produce test data for you. In this example, we combine the calling of an external feature with a data table. That means that we do not create multiple scenarios from a scenario outline but one single scenario using more complex data.

Here is `animal-title.feature`:

```
@ignore @report=false
Feature: Produce data
    Scenario: Get animals
        * def title = name + ' the ' + animal
```

This is a helper feature that should not be called on its own so, as we have seen before, we tag it with `@ignore` and `@report=false` again.

Its purpose is to create a full title from a name and an animal type; for example, if we pass `animal` with the value of `cat` and `name` with the value of `Ginny`, it should produce the title `Ginny the cat`. You can see that this feature just assumes that the `name` and `animal` variables exist. Also, it sets a new variable, `title`, that we want to use in our calling scenario.

Now, let's move to this scenario:

```
Scenario: Feature in variable
    * table animals
        | animal | name    |
        | 'cat'  | 'Ginny' |
        | 'dog'  | 'Eamon' |
    * def animalsWithTitles = call read('animal-title.feature')
animals
    * print animalsWithTitles
    * def onlyTitles = $animalsWithTitles[*].title
    * print onlyTitles
```

There are some specific steps that happen here:

1. A data table with `animal` and `name` columns is created and stored in the `animals` variable. This is automatically converted into JSON by Karate.

2. We can read and call our `animal-title.feature` file and pass the `animals` variable to it as a parameter by just writing it after the `read` operation. The `animal` and `name` variables are then automatically available within it.

3. Inside of animal-title.feature, the title variable is created from animal and name. The title variable is automatically added to the passed JSON structure. Since we use call read here, the complete return value from the called feature can be assigned to a new variable, animalsWithTitles.

    Printing this variable shows us the new JSON structure:

    ```
    [
        {
            "name": "Ginny",
            "title": "Ginny the cat",
            "animal": "cat"
        },
        {
            "name": "Eamon",
            "title": "Eamon the dog",
            "animal": "dog"
        }
    ]
    ```

4. Finally, we use JSONPath to extract only the titles from the animalsWithTitles variable into an array:

    ```
    def onlyTitles = $animalsWithTitles[*].title
    ```

5. As the result, this scenario gives us the output of the animals' titles:

    ```
    [
        "Ginny the cat",
        "Eamon the dog"
    ]
    ```

This is a more complex example but hopefully, it demonstrates how flexible data creation can be. There are even more ways to dynamically retrieve and manipulate test data. Also, all these approaches can be combined, so you have a big toolbox at your disposal!

Using setup scenarios

A newer way for Karate from version 1.3.0 onward is the use of so-called **setup scenarios**. These are specially tagged scenarios that contain setup or data creation code that can be called and used in other scenarios:

```
@setup
Scenario:
    * def animals = read('animals.json')
```

This one reads our already known `animals.json` file and stores it in the `animals` variable. It is tagged with @setup, which means that this is treated like an ignored scenario, so it cannot be called on its own.

Other scenarios can now use this, for example, in a table like this:

```
Scenario Outline: Data from file
    * print name, 'is', color
    Examples:
        | karate.setup().animals |
```

With the special `karate.setup()` method, we can access all variables from @setup scenarios. In this case, we use the JSON from the setup scenario's `animals` variable to fill our example table.

This leads to the following output that we already know from the previous examples:

```
23:14:58.675 [pool-1-thread-2] INFO  com.intuit.karate -
[print] Eamon is red
23:14:58.709 [pool-1-thread-1] INFO  com.intuit.karate -
[print] Ginny is black
```

There is also a `karate.setupOnce` method that makes sure that the setup scenario is only executed one time only if it is used in multiple scenarios or in parallel runs.

Setting up data in karate-config.js

You can also call feature files directly from `karate-config.js`. This enables you to set up data before any tests run.

Let's look at the `karate-config.js` file that uses our previous `animal-title.feature` file:

```
function fn() {
  var tweety = { name: "Tweety", animal: "bird" }
  var config = {
    bird: karate.call('animal-title.feature', tweety)
  }
  return config;
}
```

We initialize the `tweety` variable with the correct JSON format that the feature file expects. Using `karate.call`, we can invoke it. Like the `call` operation we used before from the inside of our scenarios, this too accepts optional parameters to be passed to the feature file. The rest is as we already know it: the title `Tweety the bird` is added to the passed JSON and this is then made available as the global `bird` variable, so scenarios can use it:

```
Scenario: Read from karate-config.js
    * print bird.title
```

This gives us the expected output:

```
12:11:13.925 [pool-1-thread-1] INFO   com.intuit.karate -
[print] Tweety the bird
```

As a last side note, if you run tests in parallel, you might not want your data initialization to be called multiple times. In this case, you can replace `karate.call` with `karate.callonce` – this makes sure the initialization is invoked one single time and the result is cached.

Summary

In this chapter, we saw how we can implement Karate hooks to react to events and add custom code. In the same vein, we integrated Java code into Karate scenarios so that we can use its power when we need it. Also, we explored different options for keeping your tests concise and compact using mechanisms such as code reuse, calling other scenarios, and data-driven tests.

In *Chapter 8, Karate in Docker and CI/CD Pipelines*, we will look at different approaches to run Karate on build servers so that tests can be integrated within the real software development life cycle.

8
Karate in Docker and CI/CD pipelines

As we have seen in the previous chapters, Karate is very powerful for API testing. However, so far, we have only run tests on our local systems. Also, these tests were run by us on an ad hoc basis. These are very helpful while developing applications or running tests when we want some information about the current state of APIs.

In many cases, this is not enough. Instead, these tests should be run automatically within build pipelines that typically pull the current application code and build and deploy it to a test server, test it, and continue deployment until this reaches live instances to be used in production. This process is called **continuous integration/continuous deployment** (**CI/CD**), with the goal of a completely automated build and test flow.

In any case, this must be all triggered by automation, so we don't forget to run tests at the appropriate time. This does not mean, though, that we should not be informed about the test results if there are any problems. We could even go so far as to stop further deployment to live servers in case of issues on our test instances. Also, it is helpful if our Karate tests are run regularly against live instances of an application so we can use it for monitoring purposes.

> **Shift left testing**
>
> In recent times, the concept of shifting left has gained traction. This means that tests should be run at the earliest time possible. In the case of API tests, that means that they should be run as soon as a new version of an API is available on a test server. This way, it is much easier to track down what has caused issues when tests fail. Also, the correct team dealing with API development can be notified automatically and directly without delay. If all tests were run only after every part of a system was built and deployed, it would take much longer to determine which team is responsible for fixing bugs and what component caused the error.

This is usually done is by specialized build servers such as **Jenkins** or online platforms such as **GitHub** that offer this functionality.

In this chapter, we will explore different ways to run our test to suit these platforms and fulfill our requirements. It is not possible in this limited scope to give you a full step-by-step guide on how to reach a complete CI/CD setup. Rather, it is intended as a starting point to provide ideas and further learning on how to achieve this goal.

The main topics covered in this chapter are the following:

- Triggering Karate tests from shell scripts

- Running Karate tests in a Docker container

- Customizing our tests

- Integrating Karate tests into GitHub workflows

Technical requirements

The code examples for this chapter can be found at `https://github.com/PacktPublishing/Writing-API-Tests-with-Karate/tree/main/chapter08`. We will use this project as well as its GitHub repository for demonstration purposes in this chapter. This project is a clone of the database example from *Chapter 7*, as this illustrates some concepts that will be important in a while.

You will require the following:

- The system and **integrated development environment** (**IDE**) setup we completed in *Chapter 2, Setting Up Your Karate Project*.

- **Git** for accessing Git resources and using the **GitBash** shell in Windows. It can be downloaded for all **operating systems** (**OSs**) at `https://git-scm.com/downloads`.

- An optional GitHub user account to play with workflows. You can sign up using the `https://github.com/signup` URL.

  Optionally, you can install Docker Desktop to run Docker on your local system. It can be downloaded and configured here: `https://www.docker.com/get-started`. Please note that this is not required for understanding this example and could be complex depending on your OS, processor, and so on.

So far, when running Karate tests from the command line, we have used the `mvn clean test` Maven command along with some parameters. This is perfectly fine and works well. However, it must be documented correctly that this is how to run it if other teams want to trigger those tests. Also, it requires that the test code is available on a local system and that this system has Maven and Java installed. In the following chapters, we will go step by step toward a solution that does not need all this and makes it easier to run, reuse, and set up tests for third parties.

We will first look at how to trigger Karate tests from shell scripts – the first step to a successful CI/CD integration.

Triggering Karate tests from shell scripts

Most of the time, build servers run on Linux, so it is a good idea to use **Bash**, the default shell for most Linux distributions. Also, macOS has this shell built in. For Windows, it is another story. Here, the default shells are *Command Prompt* and *PowerShell*. Neither is compatible with Bash.

Creating a batch script for Windows

If you want to create a simple script for Windows Command Prompt (also known as a **batch file**) to trigger Karate tests, this is how.

In the root directory of your Karate project, create a file with a `.bat` ending, such as `run-tests.bat`. This should contain the usual Maven command to run tests from the command line:

```
mvn clean test
```

This is the absolute bare-bones command we need for the Karate run. We could now run it from Windows Command Prompt outside of **Visual Studio** (**VS**) Code. For our purpose, let's use VS Code's terminal window. Here, you need to make sure that it shows the **cmd** icon and not **powershell** or **bash** to use it with Command Prompt.

Figure 8.1 – Running the batch file in VS Code's terminal

If it shows the wrong icon, you can click on the plus icon right next to it and open the correct terminal window:

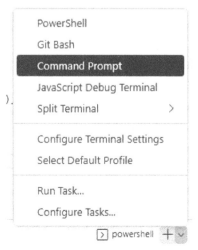

Figure 8.2 – Selecting Command Prompt

If you now run the `run-tests.bat` file, you should see that VS Code switches to a Java window and executes the Maven command that runs our tests.

This might not seem to be a big improvement. However, there could be much more code in the batch file, such as spinning up servers, adding test data to a database, and so on. The user wanting to just run tests does not have to remember to do all this by hand but just to run the provided batch file.

> **Running a batch file in PowerShell**
>
> If you want to run the same batch file in a PowerShell window, you need to invoke it by typing `.\run-tests.bat`. This signals to PowerShell that this batch script can be trusted. You will see a similar mechanism in Bash scripts.

Let's now do this as a Bash script that we can eventually run on a build server.

Creating a Bash script

A Bash script has the `.sh` file ending and should start with a header line to indicate which shell this is supposed to run on. So, our file will be called `run-tests.sh` and looks like this:

```
#!/bin/sh
mvn clean test
```

The `#!` is called a **shebang** and the path afterward points to the shell that should be used. The standard path of Bash is `/bin/sh`. The rest of this command is the same for now.

If you are using Linux or macOS, you can open a terminal and invoke this script by typing the following command:

```
./run-tests.sh
```

Depending on the system and terminal you use, you might get an error message like this:

```
permission denied: ./run-tests.sh
```

This is due to the permissions of this file. If this happens, make sure that this file has the correct permission so that it can be executed by typing the following and pressing *Enter*:

```
chmod +x run-tests.sh
```

The `chmod` command (short for *change mode*) can be used to alter file permissions. With the +x option, we specify that it should be executable. After this, running the file should be possible without problems.

Running the Bash script in VS Code on Windows

If you are on Windows and have installed Git as described in this chapter's technical requirements, you can switch VS Code to **Git Bash**. This is an emulation of Bash for Windows that behaves like the real one.

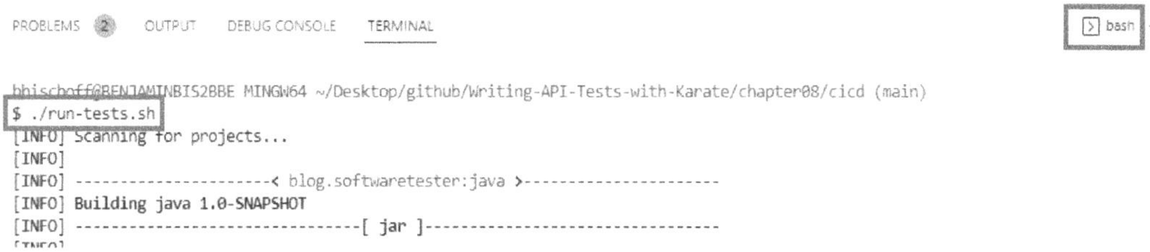

```
PROBLEMS  2    OUTPUT    DEBUG CONSOLE    TERMINAL                                      >  bash

bhischoff@BENJAMINBIS2BBE MINGW64 ~/Desktop/github/Writing-API-Tests-with-Karate/chapter08/cicd (main)
$ ./run-tests.sh
[INFO] Scanning for projects...
[INFO]
[INFO] --------------------< blog.softwaretester:java >--------------------
[INFO] Building java 1.0-SNAPSHOT
[INFO] --------------------------[ jar ]--------------------------
[INFO]
```

Figure 8.3 – Running a shell script in a VS Code Git Bash terminal

This makes it possible to test shell scripts in Windows without too much hassle.

For now, we will keep working with the Bash script, as this is what will eventually be used on a build server. Windows-based build servers are just too rare to find in the wild to spend more time on them at this point.

Let's look at the next piece of our puzzle: how to run tests inside a Docker container.

Running Karate tests in a Docker container

Often, build servers don't have all the technology necessary to run every software and every testing framework without any further configuration. Unfortunately, you don't always have complete administrator access to simply add missing components or even make complete configuration changes.

To recap, our Karate tests have some requirements they need to run properly: Java and Maven. There might be even more needed dependencies, depending on your test scope, which would require even more effort to provide.

This is where Docker comes in.

Understanding Docker

Docker is a solution that can run on many platforms. It uses a concept called **OS-level virtualization**. That means that we can have multiple isolated containers on an OS, including all kinds of custom tools and settings. For programs that run inside such a container, it feels like a self-contained system. The great thing about it is that we can have ready-made containers that include all the required resources needed to run a specific command without having to install everything individually by the user. All that is needed in that case is that the user or machine has Docker.

As with some other topics in this book, Docker is something that cannot be fully discussed in a few pages. If you want to know more about virtualization and Docker, I strongly encourage you to read more about it in the official Docker documentation at `https://docs.docker.com`.

Installing Docker

We need to have Docker installed on our local system to try out our dockerized runs. The easiest way to get started is via **Docker Desktop**. This is available at `https://www.docker.com/products/docker-desktop` for all major OSs.

Certain prerequisites are necessary depending on your OS. You can find all the required documentation here: `https://www.docker.com/get-started`. Unfortunately, the installation for Windows is not as straightforward as for a Linux or macOS system. For example, it requires **Windows Linux Subsystem (WSL)**.

> **More about WSL**
>
> There are multiple ways to run Linux commands on Windows. The most versatile one is WSL. This makes it possible to install a complete Linux distribution that can be used right from Windows without having to have a virtual machine or to switch back and forth between systems. Since it is a complete Linux, you have all the command line tools available that a typical Linux-based server has, so this is a recommended way if you do regular development on Windows that is aimed at Linux and Unix machines. Also, Docker needs this to run correctly on Windows.
>
> More about the installation and usage of WSL can be found at `https://learn.microsoft.com/en-us/windows/wsl/install`.

For our demonstration, having Docker installed locally is not required as the final Docker container will eventually run within GitHub and not on your local system. If you want to follow along, you can, of course, set it up locally as well.

Starting and verifying a Docker installation

After starting **Docker Desktop**, it will take a while until the **Docker Engine** starts. While it is doing that, you will see the message shown in the following screenshot:

Docker Desktop starting...

Figure 8.4 – Docker Desktop start up

If this says Docker Desktop Stopped or just closes, there might be a deeper problem on your system that prevents it from starting, such as a missing WSL installation or insufficient rights.

Eventually, it should look like this:

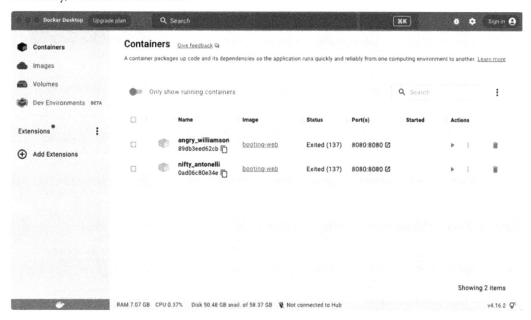

Figure 8.5 – The Docker user interface after starting up

We can also verify it is correctly running by opening a terminal window and typing `docker -v`. This should return the current Docker version like this:

```
Docker version 20.10.22, build 3a2c30b
```

Downloading a Docker image

Our goal is to run the tests regardless of the system having Maven and Java installed. Also, we want to be in control of the Maven version so we can be 100% sure that the tests are running as expected. For this purpose, there are ready-made Docker container images that are directly downloadable and usable without additional preparation. In the case of the Maven Docker images, this means that Docker can run a container with this image on any system that has Docker installed. This image then provides Maven but only if the container is running.

The official Maven container images are available on **Docker Hub**, Docker's central repository of containers. If you look at `https://hub.docker.com/_/maven`, you will see the available images. These are typically identified by tags that include the name and version of the software that this image includes:

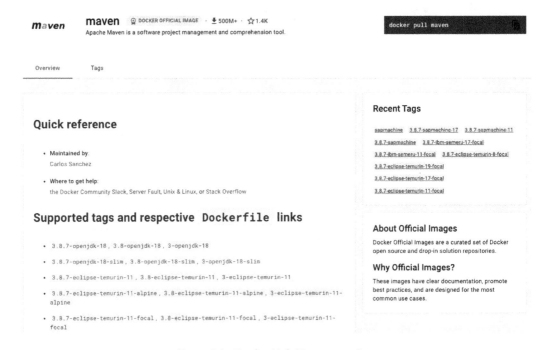

Figure 8.6 – Docker Hub Maven overview

If we did not care about the concrete version of Maven, we could just use the `docker pull maven` command, which downloads the latest Maven container by default. This is also shown on the top right of Docker Hub, as shown in *Figure 8.6*:

```
Using default tag: latest
latest: Pulling from library/maven
10ac4908093d: Pull complete
```

```
6df15e605e38: Pull complete
...
Digest: sha256:76789e7bf6713b7fe617b0e72ccf1e0cc23425bc41610c87
8f13a9b2ffd2127d
Status: Downloaded newer image for maven:latest
docker.io/library/maven:latest
```

This means we now have this container on our system, but *we don't use it yet*.

You can also verify this, either by typing `docker images` in the terminal window or by switching to the **Images** tab in Docker Desktop:

☐ **maven** ad1181366eca ⎘	latest	Unused	2 days ago	535.16 MB

Figure 8.7 – Downloaded Maven image

The nice thing is that this is only done once if the version is up to date.

For our demo purposes, we will use a specific Maven version. I will use Maven version *3.8.6*, which I also have installed on my local system. If we just add this to the command like this, we get what we want: `docker pull maven:3.8.6`.

Running a Maven command in a Maven Docker container

There are multiple ways to run commands within a Docker container. Let's run a command and then break down exactly what's going on here.

This should be run in Git Bash if you are on Windows:

```
docker run --rm --name karate-tests -v "$(pwd)":/usr/src/
mymaven -w /usr/src/mymaven maven:3.8.6 mvn -v
```

On Windows, there can be an issue with invalid paths where you get the following error (xxx refers to a directory on your local system):

```
docker: Error response from daemon: the working directory 'xxx'
is invalid
```

If this is the case, you should add the `MSYS_NO_PATHCONV=1` environment variable like this:

```
MSYS_NO_PATHCONV=1 docker run --rm --name karate-tests -v
"$(pwd)":/usr/src/mymaven -w /usr/src/mymaven maven:3.8.6 mvn
-v
```

This enables using relative paths on Windows.

A good tip is that you can also break down Docker commands into multiple lines using a backslash as a line end like this:

```
docker run \
    --rm \
    --name karate-tests \
    -v "$(pwd)":/usr/src/mymaven \
    -w /usr/src/mymaven \
    maven:3.8.6 \
    mvn -v
```

It is very important that there is no space after the backslashes, they need to be followed by line breaks directly.

Let's walk through this line by line:

1. With `docker run`, we can run a container that is isolated from the host system. So, by default, programs running inside the container don't have access to the host system unless it's specifically wanted.

2. The `--rm` option removes the container as soon as the commands running inside it stop. This means that the container only runs if it is needed.

3. With `--name`, we can give it a name so we can identify it while it is running, for example, within the Docker Desktop **Containers** tab.

4. The next line is a little more complex. The `-v` can specify a directory or a file from the host system and make it accessible from inside the container. Here, we take the current directory where the Docker command is running using the `$(pwd)` command. After `:`, we can specify what this directory should be called inside the container. In this case, it is `/usr/src/mymaven`. Essentially, we are building a bridge to a directory on the host system. This is called **mounting**.

5. With `-w`, we now take the mounted directory from before and make it the working directory of the Docker container. That means that the container now goes to the same directory as the host system and executes all coming commands inside this directory.

6. This is followed by the name of the container image we want to use, which is our Maven image with the `maven:3.8.6` tag.

7. Finally, the Maven container accepts a Maven command as a parameter. Here, we use `mvn -v`, which displays the current Maven version. Note that this should display `3.8.6` regardless of which Maven version is installed on the host system because it is the Maven version that is available inside the running container.

Running this command displays the expected output:

```
Apache Maven 3.8.6 (84538c9988a25aec085021c365c560670ad80f63)
Maven home: /usr/share/maven
Java version: 17.0.5, vendor: Eclipse Adoptium, runtime: /opt/
java/openjdk
Default locale: en_US, platform encoding: UTF-8
OS name: "linux", version: "5.15.49-linuxkit", arch: "amd64",
family: "unix"
```

If you want, you can test the same Docker command with the old maven:3.3.9 container. In this case, the Maven version is correctly reported as 3.3.9.

Running Karate tests inside a Maven Docker container

In the last section, we used mvn -v as a sample command to test out our Maven container. Perhaps you can already see what this is all about since we also trigger Karate via Maven; we can also just give the command for it to the container. However, this only works if we call the docker run command from the directory where the Karate tests pom.xml file is located:

```
docker run \
    --rm --name karate-tests \
    -v"$(pwd)":/usr/src/mymaven \
    -w /usr/src/mymaven \
    maven:3.8.6 \
    mvn clean test
```

This runs the Maven container using the mvn clean test command that triggers our Karate tests. Since this is connected to the host system, we can even find our test report under the target directory there as if we had run the tests without Docker!

Running our shell script inside Docker

Let's get back to our previous little shell script. Remember that this is ultimately what we want to use because we can add much more functionality to it in the future. We are already mounting the current directory from which we start the Docker command. This directory also hosts our shell script. So, instead of the mvn clean test command from before, we can now run the shell script like this:

```
docker run \
    --rm --name karate-tests \
    -v"$(pwd)":/usr/src/mymaven \
```

```
-w /usr/src/mymaven \
maven:3.8.6 \
./run-tests.sh
```

This runs our Karate tests as expected within the Maven container. While the tests are running, we can see our Maven Docker container appearing in Docker Desktop as well:

Figure 8.8 – Running the karate-tests Maven 3.8.6 container

We can also see that all Maven dependencies are downloaded before the tests start.

```
Downloading from central: https://repo.maven.apache.org/maven2/org/apache/maven/
surefire/surefire-junit-platform/2.22.2/surefire-junit-platform-2.22.2.pom
Downloaded from central: https://repo.maven.apache.org/maven2/org/apache/maven/s
urefire/surefire-junit-platform/2.22.2/surefire-junit-platform-2.22.2.pom (7.0 k
B at 194 kB/s)
Downloading from central: https://repo.maven.apache.org/maven2/org/apache/maven/
surefire/surefire-providers/2.22.2/surefire-providers-2.22.2.pom
Downloaded from central: https://repo.maven.apache.org/maven2/org/apache/maven/s
urefire/surefire-providers/2.22.2/surefire-providers-2.22.2.pom (2.5 kB at 71 kB
/s)
Downloading from central: https://repo.maven.apache.org/maven2/org/junit/platfor
m/junit-platform-launcher/1.3.1/junit-platform-launcher-1.3.1.pom
Downloaded from central: https://repo.maven.apache.org/maven2/org/junit/platform
/junit-platform-launcher/1.3.1/junit-platform-launcher-1.3.1.pom (2.2 kB at 69 k
B/s)
Downloading from central: https://repo.maven.apache.org/maven2/org/apiguardian/a
piguardian-api/1.0.0/apiguardian-api-1.0.0.pom
Downloaded from central: https://repo.maven.apache.org/maven2/org/apiguardian/ap
iguardian-api/1.0.0/apiguardian-api-1.0.0.pom (1.2 kB at 32 kB/s)
Downloading from central: https://repo.maven.apache.org/maven2/org/junit/platfor
m/junit-platform-engine/1.3.1/junit-platform-engine-1.3.1.pom
Downloaded from central: https://repo.maven.apache.org/maven2/org/junit/platform
/junit-platform-engine/1.3.1/junit-platform-engine-1.3.1.pom (2.4 kB at 75 kB/s)
Downloading from central: https://repo.maven.apache.org/maven2/org/junit/platfor
```

Figure 8.9 – Maven dependencies download when running in Docker

This is because the Maven Docker container does not contain these dependencies when it starts up and must fetch them again every time it is spun up. We can suppress this endless list of logs by calling the Maven command with the --no-transfer-progress flag, so it looks like this in run-tests.sh:

```
mvn clean test --no-transfer-progress
```

Another issue is that you cannot stop the container when pressing *Ctrl + C* as you would expect in a shell. To fix that, we can add the `-it` flags to the Docker command:

```
docker run \
    -it --rm --name karate-tests \
    -v"$(pwd)":/usr/src/mymaven \
    -w /usr/src/mymaven \
    maven:3.8.6 \
    ./run-tests.sh
```

This makes the running container *interactive*, so it reacts to the keyboard command to terminate it. Note that this is only relevant when using the `docker run` command locally. In fact, some build servers such as Jenkins disallow the `-it` flags.

Customizing our tests

Now we took the first steps towards running Karate in a CI/CD environment. There is one severe problem that needs to be addressed, though.

Currently, the tests won't work without hardcoded database credentials. This is not what we want. Ideally, these should come from a secure place that can only be changed by administrators. In the next sections, we will prepare our tests for this scenario.

Passing database credentials as system properties

If you remember from the last chapter, to access our test database, we needed these parameters in the `MySQL.java` class:

- `connectionString`: With the host and database name.

- `user`: If you use a database from `https://www.freesqldatabase.com`, the user name is the same as the database name. For other database providers, this can be different, though.

- `password`: The password for the database.

In our source code, these parameters look like this:

```
String connectionString =
    "jdbc:mysql://sql7.freesqldatabase.com/sqlxxx";
String user = "sqlxxx";
String password = "xxx";
```

To populate these parameters with dynamic values, we must modify the `karate-config.js` file so it can provide them based on passed system properties:

```
function fn() {
  var config = {
    host: karate.properties['host'],
    db: karate.properties['db'],
    pass: karate.properties['pass']
  }
  return config;
}
```

This way, we can call the tests like this:

```
mvn clean test -Dhost=xxx -Ddb=yyy -Dpass=zzz --no-transfer-
progress
```

The values xxx, yyy, and zzz are placeholders because I don't want to disclose my personal database credentials here.

As a side note, as with the `docker run` command, we can make the Maven command more readable in the `run-tests.sh` script by using the backslash trick to break up the statement into multiple lines:

```
mvn clean test \
-Dhost=$HOST -Ddb=$DB -Dpass=$PASS \
--no-transfer-progress
```

Again, this can be problematic on a Windows system. To use it there, make sure that the `run-tests.sh` script is saved using **Line Feed (LF)** line endings and not **Carriage Return Line Feed (CRLF)**. To do this in VS Code, click on **CRLF** in the footer:

Figure 8.10 – Triggering the line endings drop-down box in VS Code

After that, you can choose **LF** in the drop-down box that is automatically triggered:

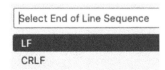

Figure 8.11 – Choosing LF line endings in VS Code

Now, we need to alter the database connection code to use these system properties.

Using the system properties in the database connection class

Let's modify the MySQL.java class to pass the credentials in the constructor and store these as private member variables. This way, we can reuse them in other methods within the same class:

```
private final String host;
private final String db;
private final String pass;

public MySQL(final String host,
             final String db,
             final String pass) {
    this.host = host;
    this.db = db;
    this.pass = pass;
}
```

Now we can use these variables instead of the hard-coded strings from before. This is the modified portion of the getMagicians method, where we now construct the JDBC connectionString variable using the new host and db variables. Also, in the getConnection method, we now use db instead of the hard-coded username from before and pass in place of the hard-coded password:

```
public String getMagicians() throws Exception {
    String connectionString =
        "jdbc:mysql://" + host + "/" + db;
    try (
        Connection c = DriverManager.getConnection(
            connectionString, db, pass
        );
        Statement s = c.createStatement();
        ResultSet result =
            s.executeQuery("select * from magicians")
    ) {

        ...

    }
}
```

When passing the correct credentials on the command line, the database connection should be established correctly, and the test should pass again.

Let's see how we can achieve the same when running the Maven command within Docker.

Passing parameters from Docker

Since we cannot easily pass system properties to a Docker container, we must find another way to parameterize our run. Luckily, the docker run command has the option to pass environment variables to the running container using the -e option. This is a mechanism we can use for our purpose:

```
docker run \
    -e HOST=xxx -e DB=xxx -e PASS=zzz \
    -it --rm --name karate-tests --init \
    -v"$(pwd)":/usr/src/mymaven \
    -w /usr/src/mymaven \
    maven:3.8.6 \
    ./run-tests.sh
```

Here, we set three environment variables, HOST, DB, and PASS – writing them in all uppercase is the general convention. Note that here I am using placeholder values (xxx, yyy, and zzz) for demonstration purposes.

Now we must access these within our run-tests.sh shell script so we can pass them along to Karate as system properties.

Environment variables

Environment variables are accessible from shell scripts by preceding them with a $ sign:

```
mvn clean test -Dhost=$HOST -Ddb=$DB -Dpass=$PASS
```

We already know that we can pass system variables to Maven using the -D option. So, essentially, what we are doing here is taking the environment variables passed to Docker and relaying them to Maven as system properties.

> **System properties versus environment variables**
>
> System properties are key-value pairs that can be set and retrieved by Java at any point. Environment variables, on the other hand, are key-value pairs as well, but they are available on the OS level. This means they can be read by all programs that run on the same computer. Java cannot set environment variables, only system properties.

If you test this now with your real credentials in the `docker run` command, you should see that the test is still working as before. Now, however, we made a lot of progress in passing the required settings from the outside. As we will see next, there is yet another layer, so we won't have to hardcode these parameters into our `docker run` command when this is running in a CI/CD pipeline.

We will see in the next section how GitHub workflows are used and how we can integrate what we already implemented into a real pipeline.

Integrating Karate tests into GitHub workflows

In this part of the chapter, we will use the popular GitHub workflows to integrate Karate tests into a real build pipeline. We will do this both with and without Docker to see the differences.

Understanding GitHub workflows

If you do development or code reviews, chances are that you regularly use GitHub already. This Microsoft-owned Git-based code management platform is very popular among development teams. It offers pretty much all the required features that are needed within the software development life cycle.

Among many others, it includes version control, bug tracking, and – quite important for this chapter – CI and CD functionality. This is integrated into the platform by so-called GitHub actions that allow building such pipelines without third-party dependencies. These pipelines can be triggered manually or react to various events such as code commits, pull requests, branch creation, modification, deletion, and many more. Apart from these automatic events (also called triggers), GitHub workflows can also be configured to be started manually or on a fixed schedule.

Workflows are stored in the `.github/workflows` directory of your repository. They are YAML files ending with `.yml` that define what should be executed where, in which order, and in reaction to what.

Inside workflows, you define jobs that contain specific steps to be executed. These steps can run either predefined actions or completely custom scripts. We will use both.

In the following sections, we will set up a demo workflow to showcase how Karate tests can be integrated. First, we deal with another problem, though.

Managing secrets

When defining a workflow, we need to provide the database credentials for our test. We could provide them when triggering the workflow, as GitHub also supports using parameters when starting a workflow manually. However, this would mean remembering and entering this data every time we want to run the test job. Also, it would defeat the purpose of having a fully automated build pipeline when the test step requires manual intervention to enter data.

Like all major build servers, GitHub has a built-in **secrets management** that we can use to solve this. That means that we can store important information in encrypted form and use this in our workflow files. This is perfect for our use case.

To enter secrets, go to **Settings | Secrets and variables | Actions**. You need to have the appropriate rights to manage secrets, so if you don't see this, you were not given permission by the repository administrator to manage secrets. If you do have permission, you can click on the **New repository secret** button and enter a key and a value:

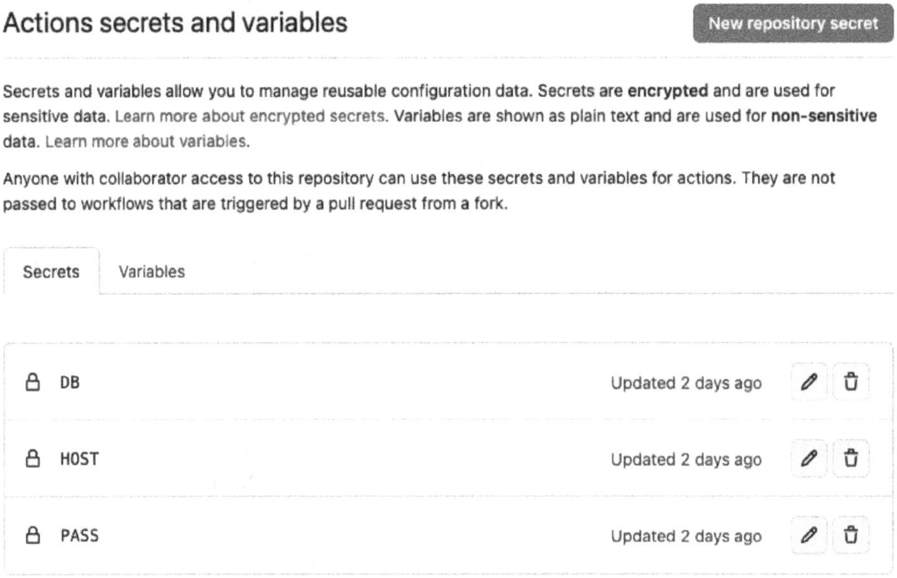

Actions secrets and variables

New repository secret

Secrets and variables allow you to manage reusable configuration data. Secrets are **encrypted** and are used for sensitive data. Learn more about encrypted secrets. Variables are shown as plain text and are used for **non-sensitive** data. Learn more about variables.

Anyone with collaborator access to this repository can use these secrets and variables for actions. They are not passed to workflows that are triggered by a pull request from a fork.

Secrets	Variables

🔒 DB	Updated 2 days ago	✏️ 🗑️
🔒 HOST	Updated 2 days ago	✏️ 🗑️
🔒 PASS	Updated 2 days ago	✏️ 🗑️

Figure 8.12 – GitHub secrets management

Here, I added three secrets: **DB**, **HOST**, and **PASS** with my `freesqldatabase.com` credentials. Be careful; once the secrets are entered and saved, you will not be able to see this information anymore. The only available options are to delete them or update their respective values. In the following code examples, we can access these secrets using GitHub's built-in `secrets` map, as we will see next.

Adding a GitHub workflow to a repository

In the GitHub repository of this book, I defined two workflow files: `run-karate-docker.yml`, which runs the tests using Docker, and `run-karate-maven.yml`, which uses pure Maven.

Figure 8.13 – Workflow YAML files

These workflows can be seen in GitHub when going to the **Actions** tab in the repository that includes the YAML files:

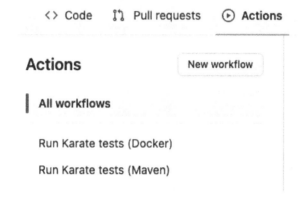

Figure 8.14 – Actions overview in GitHub

Let's look at the first workflow called **Run Karate tests (Docker)**.

Setting up the Docker-based GitHub workflow

We will now look at how to put our knowledge of Docker to use. Following that, we will write a GitHub workflow that can be triggered manually and runs our Karate tests within Docker.

> **GitHub triggers**
>
> Usually, this kind of workflow would be triggered by new code being pushed to the repository or any other preceding actions, but this would make our demo harder to follow. That's why I chose a manual trigger for this example. Once you understand the concept, you can potentially incorporate this into real, more complex project pipelines on GitHub.
>
> The official GitHub workflow documentation at `https://docs.github.com/de/actions/using-workflows/events-that-trigger-workflows` has more information about the different kinds of triggers.

This is the complete code for our GitHub workflow stored in `./github/workflows/run-karate-docker.yml`:

```yaml
name: Run Karate tests (Docker)

on:
  workflow_dispatch:

jobs:
  karate-tests:
    runs-on: ubuntu-latest
    steps:
    - uses: actions/checkout@v3
    - name: Run with Docker
      working-directory: chapter08/cicd
      run: >
        docker run \
        -e HOST=${{ secrets.HOST }} \
         -e DB=${{ secrets.DB }} \
         -e PASS=${{ secrets.PASS }} \
        --rm --name karate-tests --init \
        -v"$(pwd)":/usr/src/mymaven \
        -w /usr/src/mymaven \
        maven:3.8.6 \
        ./run-tests.sh
```

It is important to have the correct spacing in these YAML files, or we could break them. In case of errors, GitHub will notify us about them.

Let's find out what this code does:

1. We start with `name:  Run Karate tests  (Docker)`, which defines the name that is shown within the GitHub **user interface (UI)**.

2. Next, we define how this workflow can be triggered using `on`. In our case, we want a manual trigger called `workflow_dispatch` in GitHub:

    ```
    on:
       workflow_dispatch:
    ```

    The `:` symbol afterward is correct since the manual trigger could have more options that could follow it, such as the definition of parameters. In our case, we don't need this, so we have nothing after `:`.

3. Now, we tell GitHub what our jobs are:

    ```
    jobs:
       karate-tests:
    ```

4. Here, we only have one job named `karate-tests`. Again, the `:` symbol means that there are more settings below.

5. The next line specifies which OS the `karate-tests` job should run on. The `runs-on: ubuntu-latest` setting is a good choice since it is a Linux system that comes with Docker preinstalled. If you want to know more about the OS options and what software is preinstalled, check out the official documentation here: `https://docs.github.com/en/actions/ using-github-hosted-runners/about-github-hosted-runners`.

6. Now, we start specifying the steps that belong to our `karate-tests` job. This is done with the `steps:` label. Naturally, we first need to give the workflow access to the main branch of our repository so that it can run the Karate test code later:

    ```
    steps:
       - uses: actions/checkout@v3
    ```

    You can see that we specify `actions/checkout` in version *3* (`@v3` adds the version), which does what we want. Actions, as mentioned in the introduction to this chapter, are predefined plugins for GitHub workflows that serve specific purposes such as checking out branches, sending messages, or connecting to various other systems.

7. After we have the code checked out, we need to go to the directory where our Karate tests are. With `working-directory: chapter08/cicd`, we tell the GitHub workflow to do exactly that (the tests we want to run are in the `chapter08/cicd` directory).

8. Now that we are in the right directory, we can finally call the Docker command that we developed earlier. To do that, we use GitHub's `run` command like this:

```
run: >
    docker run \
        -e HOST=${{ secrets.HOST }} \
        -e DB=${{ secrets.DB }} \
        -e PASS=${{ secrets.PASS }} \
        --rm --name karate-tests --init \
        -v"$(pwd)":/usr/src/mymaven \
        -w /usr/src/mymaven \
        maven:3.8.6 \
        ./run-tests.sh
```

You can see that we pass the secrets that are stored in GitHub instead of the hard-coded database credentials.

Running the workflow

We created the workflow with a `workflow-dispatch` trigger so we can easily start this from the GitHub UI.

In the **Actions** tab, you can select the workflow to run, in this case, **Run Karate tests (Docker)**. With a click on **Run workflow**, a menu opens where you can select a different branch except **main**. This is beneficial if tests change within a specific branch, and those should be verified before merging this to the main branch. If our workflow had parameters defined, those could also be set in this menu. With a click on **Run workflow**, GitHub will start it with the specified `ubuntu-latest` runner:

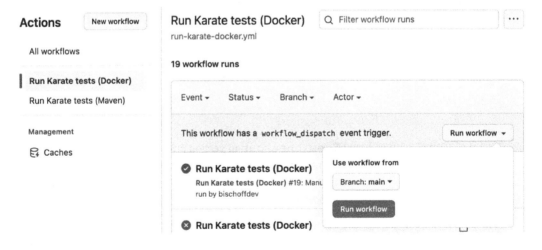

Figure 8.15 – Manually triggering a workflow on GitHub

This is how the workflow looks while it is running:

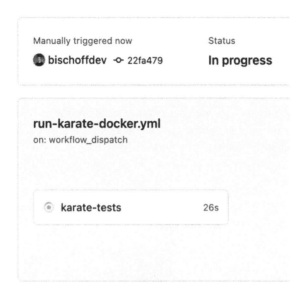

Figure 8.16 – A running GitHub workflow

This view shows the workflow and job names as well as their status and runtime. Also, you can see that it was triggered manually and by whom.

In the end, it should look like this if all goes well:

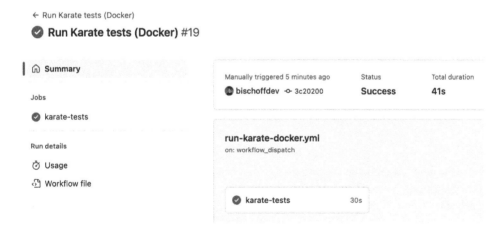

Figure 8.17 – Successful workflow run

Let's add the test report as the final step.

Adding the test report

GitHub can store any files within a workflow which is a great feature for test reports that should be available within each test run. You can read more about it at `https://docs.github.com/en/` `actions/using-workflows/storing-workflow-data-as-artifacts`.

Unfortunately, when saving these, they are zipped and attached to the respective workflow run as a download. There are ways around this, such as storing the reports outside of GitHub or using GitHub actions that generate compatible reports from Surefire XML files, for example (we covered these in *Chapter 5, Reporting and Logging*). However, this would go too far here, and for this demo, we'll settle for the packaged version.

For this example, I added the `Cluecumber` report dependency that we discussed in *Chapter 5* to the project and modified the runner class like this:

```
@Test
void testParallel() throws CluecumberException {
    Results = Runner.path("classpath:examples")
        .outputCucumberJson(true)
        .reportDir("target/myReport")
        .outputHtmlReport(false)
        .parallel(1);

    new CluecumberCore.Builder().build()
        .generateReports(
            "target/myReport",
            "target/customReport");

    assertEquals(0, results.getFailCount(),
        results.getErrorMessages());
}
```

This creates the final test report in the `target/customReport` directory. Now we can add some code to the bottom of the `run-karate-docker.yml` GitHub workflow file to store this report within each run:

```
- uses: actions/upload-artifact@v3
  with:
    name: Karate Report
    path: chapter08/cicd/target/customReport
```

The new step uses GitHub's `upload-artifact` action to do that. We can specify the path to our test report (`chapter08/cicd/target/customReport`) as well as a name that should be displayed within GitHub. This is all we need to do.

Running the test now, we should see that under **Artifacts**, our test report is displayed and can be downloaded by clicking on its name:

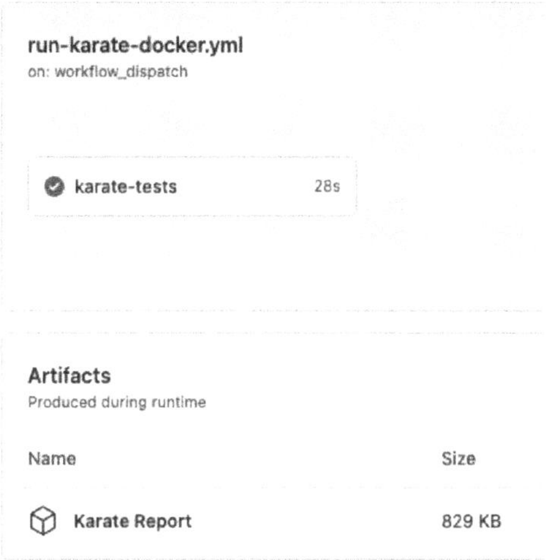

run-karate-docker.yml
on: workflow_dispatch

✓ karate-tests   28s

Artifacts
Produced during runtime

Name	Size
📦 **Karate Report**	829 KB

Figure 8.18 – The test report attached to a workflow run

Now we have the report available in each workflow run. There are, of course, many more things you could potentially do here, such as only attaching reports when tests failed, automatically sending the report to stakeholders via email or instant messaging, storing a report history, uploading reports to a web server, and many more. This topic could easily be a book in itself!

Using Karate in a GitHub workflow without Docker

Let's quickly finish this chapter with a similar workflow that does not use Docker at all. If your CI/CD system has Java and Maven installed, the non-Docker approach is generally faster, but you may be in less control of the Maven and Java versions or other installed libraries.

GitHub has Maven available in its runners, so we can add the `run-karate-maven.yml` workflow file I mentioned at the start of this chapter:

```
name: Run Karate tests (Maven)

on:
  workflow_dispatch:
```

```
jobs:
  karate-tests:
    runs-on: ubuntu-latest
    steps:
    - uses: actions/checkout@v3
    - name: Set up JDK 11
      uses: actions/setup-java@v3
      with:
        java-version: '11'
        distribution: 'temurin'
        cache: maven
    - name: Karate tests
      working-directory: chapter08/cicd
      env:
        HOST: ${{ secrets.HOST }}
        DB: ${{ secrets.DB }}
        PASS: ${{ secrets.PASS }}
      run: ./run-tests.sh
    - uses: actions/upload-artifact@v3
      with:
        name: Karate Report
        path: chapter08/cicd/target/customReport
```

About half of the code is the same as in the Docker version. The highlighted lines set up Java and run the tests with Maven.

This is what happens there:

1. After the code checkout, we use GitHub's `setup-java` action to prepare a specific Java version that we want to use. In this case, I selected *Java 11* from the **Temurin** distribution, which is essentially a completely free and open version of the Java runtime environment (*temurin* is an anagram of *runtime*). You can read more about it here: `https://adoptium.net`. Another setting that speeds up test runs is `cache: maven`. This tells GitHub to store all Maven libraries needed for our Karate test runs in its cache, so they don't need to be re-downloaded from the Maven central repository every time.

2. Next, we run the `./run-tests.sh` script as before, with the big difference that now it is not running within Docker. To pass our GitHub secrets, we use the `env` option that sets the `HOST`, `DB`, and `PASS` environment variables that are used within the script.

The rest stays as before. Running this workflow in GitHub now produces a similar result as the Docker version:

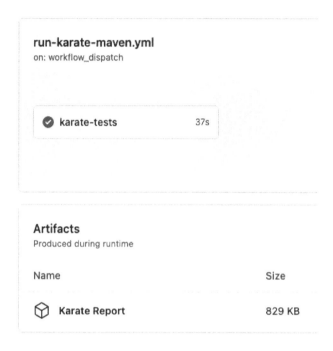

Figure 8.19 – Maven-only workflow run without Docker

However, it is much faster after the second run due to GitHub's Maven caching!

Summary

This concludes our example GitHub workflow run to showcase how we can integrate Karate into a CI/CD pipeline. For this purpose, we have addressed the writing of shell scripts, took a chance on Docker, and finally brought all the pieces of the puzzle together in a GitHub workflow. Also, we looked at how we can store reports and run our tests in GitHub workflows with native Maven.

As already mentioned, this chapter cannot, of course, provide a comprehensive introduction to this big topic but should rather show which pieces need to be put together to integrate tests into such an existing system. The exact steps may differ depending on which build server you are using, but this approach can be applied to all of them.

In *Chapter 9, Karate UI for Browser Testing*, we will look at a lesser-known part of Karate that allows you to write browser tests!

9

Karate UI for Browser Testing

All previous chapters have mainly been about API testing. This is also Karate's main field of application, as we have clearly seen. However, it is often forgotten that Karate offers much more.

We can also use it as a browser-based UI testing tool. This opens completely new areas of application for Karate and breaks the boundaries of those testing tools that are optimized for only one of these purposes. Also, this can be a great advantage for onboarding new employees when it is only necessary to familiarize them with a single test framework.

We will look at how Karate compares to other browser testing tools, how we set it up to work with multiple browsers, and, of course, we will write some tests together for a website. Finally, we'll look at the debugging capabilities of Karate UI tests and learn how we can use these tests together with Karate's mocking capabilities.

In this chapter, we will cover these main topics:

- What Karate UI is for
- Writing a basic Chrome scenario
- Finding and interacting with elements
- Debugging UI tests
- Redirecting HTTP calls

Technical requirements

The code examples for this chapter can be found in this book's GitHub repository: `https://github.com/PacktPublishing/Writing-API-Tests-with-Karate/tree/main/chapter09`.

You will require the following:

- The system and IDE setup we completed in *Chapter 2*, *Setting Up Your Karate Project*
- Google Chrome for running the tests in this chapter and using DevTools

What Karate UI is for

The purpose of **end-to-end** (**E2E**) testing starting with the UI is to get a full picture of the application from the frontend—including JavaScript and CSS frameworks, cookie handling, and so on—all the way to the backend with its microservices, data storage, caches, and APIs.

This kind of testing is usually much slower and more error prone than pure API testing but at the same time, it paints a better picture of the state of the application. Consider this: if one API does not function properly, that does not necessarily mean that the functionality of the web application is restricted. If a full browser-based E2E test has a legitimate failure, the application is likely unusable.

There are many different browser testing tools available in today's market. One of the oldest and most well-known is **Selenium** (`https://www.selenium.dev`). This project established the **W3C WebDriver protocol**, which is used to automate interactions with web browsers—for example, clicking on elements, filling out inputs, and page navigation. The WebDriver protocol is implemented by *browser driver* executables, such as **ChromeDriver** and **Microsoft Edge WebDriver**. These small applications sit in between the browser and Selenium and translate the WebDriver protocol commands to browser-specific commands.

In that sense, Selenium is not a testing tool, but merely a way to remote-control a browser to simulate user behavior. However, it is often used in the context of UI testing frameworks. Another well-known testing project based on the WebDriver protocol is **Appium** (`https://appium.io`), which is used for testing mobile applications.

For Chromium-based browsers, there is a newer alternative to the WebDriver protocol called **Chrome DevTools Protocol** (**CDP**). It can access the browser in similar ways to the WebDriver protocol but offers further possibilities such as intercepting network requests. Typically, you need to decide based on the range of browsers and devices you need for your tests if you want to use CDP or the WebDriver protocol.

Lately, a lot of new competitors such as **Cypress** (`https://www.cypress.io`) and **Playwright** (`https://playwright.dev`) have entered the browser testing scene and are gaining momentum. These aim to be modern alternatives to the Selenium approach and have their own advantages and disadvantages, depending on the use cases for which they are to be deployed. Here, it is very important to do a thorough evaluation of these tools in advance to avoid roadblocks later.

Karate UI is a special case when it comes to browser tests. Since it is so strong in the field of API testing, it can communicate with browsers via WebDriver or CDP as both rely on REST calls internally. In fact, you could even send these pure REST calls to the browsers through Karate if you desire (Karate exposes the methods for this, as explained in `https://karatelabs.github.io/karate/examples/ui-test/#webdriver-tips`, but this is a very advanced topic that is unnecessary in most tests).

In the following sections, we will look at the basic functionality of such tests to give you an overview. This can then be the starting point for further learning—for example, by studying the full Karate UI documentation at `https://karatelabs.github.io/karate/karate-core`.

It's not for nothing that there are books dedicated just to UI testing, as this topic covers a lot of aspects that go beyond API testing.

Related testing capabilities

In this chapter, we will exclusively focus on browser automation; otherwise, this book would be twice the size. Before we go on with this, I would like to mention that Karate has some more UI test-related features that I do not want to withhold here for reference:

- Karate UI can use the mentioned Appium project to test mobile applications. However, this feature is not under heavy development and is not considered quite as important. However, if you want to look at it, you can find some limited information here: `https://karatelabs.github.io/karate/examples/mobile-test`.

- Karate can even communicate with the Playwright framework via a special driver. This requires you to set up a Playwright server and might be something to consider if you want to want to transition from Playwright to Karate. More information about this can be found here: `https://karatelabs.github.io/karate/karate-core/#playwright`.

- Karate can also be used for Windows desktop application automation by building a bridge to **Robot Framework** (`https://robotframework.org`). You can read more about this integration here: `https://karatelabs.github.io/karate/karate-robot`.

With that out of the way, let's now look closer at how to set up Karate UI to use it efficiently.

Writing a basic Chrome scenario

Karate's documentation recommends starting with automating the Chrome browser since it is the easiest and most capable. So, we will now create our first basic browser test for Chrome.

Opening a website in Chrome

We need to know in advance which browser we want to use so that we can specify the correct *driver*. Based on this, Karate will use the desired browser and protocol. For this example, we will start with the most straightforward one: Chrome.

Configuring the Karate driver

We will create a new scenario called `First test` and gradually add functionality during the next sections:

```
Scenario: First test
    * configure driver = { type: 'chrome' }
```

As with other Karate-specific settings we have seen in earlier chapters, we use the `configure` keyword. This time, we tell Karate that our `driver` instance should be of type `chrome`. This `driver` instance should not be confused with *WebDriver* from the W3C WebDriver protocol. Instead, `driver` can be understood as how Karate should interact with a specific browser. The `chrome` value refers to the Chrome executable using the DevTools protocol. Karate will attempt to locate Chrome in the operating system's default location.

> **Running in other browsers**
>
> In this chapter, we exclusively work with Chrome in order not to make the introduction to browser tests unnecessarily complicated. If you want to continue later with running Karate UI tests in different browsers, it is mostly just a matter of specifying a different value for the driver type. A list of valid driver types and which protocol they use to communicate with the web browser can be seen at `https://karatelabs.github.io/karate/karate-core/#driver-types`.

If you run the test now, it should instantly pass. This happens because we're not using the driver yet but just configuring it. So, essentially, this test is doing nothing yet.

Choosing a custom executable location

If your browser installation (or WebDriver executable) is in a custom folder on your operating system, you can use this special syntax to configure the driver:

```
* configure driver = { type: 'chrome', executable: 'path_to_
executable' }
```

Here, `path_to_executable` should contain the full path to the browser or WebDriver application. If desired, you can also specify a batch file or shell script here. This is a rare case, though, and is only needed if you require a very custom setup.

Opening a website

The first step for every browser test after configuring the Karate driver is to open a specific website. When this happens, the driver that is used by Karate to communicate with the web browser is initialized and can be used further to interact with elements on the page, set cookies, and in general do everything that a user could when manually navigating and using the web application.

Opening a website is very straightforward, as shown here:

```
Given driver 'https://softwaretester.blog'
```

By just specifying a URL to open after the configured driver, the driver connects to the browser and navigates to it. Remember that the same mechanisms apply as with Karate API testing, meaning that this URL can also be taken from `karate-config.js`, helper features, system properties, and more. Also, you don't need to use the `Given`, `When`, and `Then` keywords and can replace them with the catch-all asterisk, `*`, exactly as with API tests.

When running this scenario again, you will see the browser opening, going to the URL, and quickly closing again. This happens automatically at the end of a scenario. As soon as we add more test steps, you will see the browser for a longer time. Note that this test will always pass for now since we are not interacting with anything or making matches or assertions. We will add these in the next section.

In a later section, we will explore some ways to better follow what is happening in the browser while the test is running.

The most important thing in every browser test framework is how to find and use web elements on a website. This next section deals with this topic.

Finding and interacting with elements

So far, our example test just opens a website. To turn it into a real test, there are several ways to locate elements so that we can interact with them and validate the behavior of the application or web page. This is a crucial step toward simulating real user behavior in a browser test.

> **HTML and CSS**
>
> It is very helpful when working with locators to understand the basics of HTML and CSS. This knowledge makes it easier to target elements, understand their relationship, and retrieve information from them. As this book is mainly about API testing and this chapter gives only a limited insight into the world of UI testing, it is worth taking a closer look at these topics via other sources.
>
> A nice little tutorial for both technologies can be found here: `https://www.w3.org/Style/Examples/011/firstcss.en.html`.

In the following sections, we will see how to select locators and which locators are supported by Karate UI.

Understanding locators

In the context of UI testing, locators are characteristics used to identify and locate UI elements on a web page or mobile application. They are typically used in test automation frameworks to interact with UI elements such as buttons, input fields, and drop-down menus.

Locators come in different types depending on the technology used to build the application or web page. Further on in this section, we will see the different locators you can use in Karate.

Knowing the different kinds of locators

Locators come in different flavors, each with its own set of pros and cons. Before looking at how to find a good locator, let's first see which kinds of locators Karate supports.

Using XPath

If you recall, we already saw the use of XPath for dealing with XML files in *Chapter 6, More Advanced Karate Features*. Since HTML is a special kind of XML, we can also use it in this context to select elements based on their attributes or relationships with other elements. It allows us to specify complex paths to an element, which can be helpful when the element cannot be easily identified otherwise.

The following are some important XPath commands as a refresher:

- `//`: This command selects elements anywhere in the document tree. It can be used to locate elements based on their tag name, attribute values, or relationships to other elements.

- `[@attribute]`: This command selects elements that have a specific attribute. For example, `//*[@id]` would select all elements that have an *ID* attribute.

- `[@attribute='value']`: This command selects elements that have a specific attribute value. For example, `//*[@class='btn']` would select all elements that have a class attribute with a `btn` value.

- `//tag[text()='text']`: This command selects elements that have specific text content. For example, `//h1[text()='Welcome']` would select the h1 element with the `Welcome` text content.

- `parent::`: This command selects the parent element of the current element. It can be useful for navigating through the document tree and locating related elements.

- `following-sibling::` and `preceding-sibling::`: These commands select elements that come after or before the current element in the document tree. They can be used to locate related elements or to navigate through lists or tables.

XPath locators can be complex, and the preceding list is by no means extensive. More information about XPath can be found here: `https://www.w3.org/TR/1999/REC-xpath-19991116`.

Using CSS locators

CSS locators can also be used to select and locate elements in UI testing. These selectors can select elements based on their tag name, ID, class, or other attributes. Here are some examples:

- `tagname`: This command selects elements based on their tag name. For example, `button` would select all *button* elements on the page.

- `.classname`: This command selects elements based on their class name. For example, `.btn` would select all elements with a class of `btn`.

- #id: This command selects an element based on its ID attribute. For example, #submit-button would select the element with an ID of submit-button.

- :nth-child(n): This command selects the *n*th child element inside of its parent. For example, li:nth-child(2) would select the *second* li element in a list.

- :nth-last-child(n): This command selects the *n*th child element *from the end* of its parent. For example, li:nth-last-child(2) would select the second-to-last li element in a list.

- :not(selector): This command selects elements that do not match a specified selector. For example, input:not([type='submit']) would select all input elements that are not **Submit** buttons.

- >: This command selects the direct child element of the parent element. For example, ul > li would select all li elements that are direct children of ul elements.

You can see that CSS locators are often simpler and more concise than XPath expressions, which can make them easier to use and maintain in test scripts.

Using friendly locators

Friendly locators are a newer approach to element location in UI testing that aims to make test scripts more robust and readable. Unlike XPath and CSS selectors, which rely on specific element attributes, friendly locators use the relationships to nearby elements to identify another element. You can view some examples here:

- rightOf: This locator is used to locate an element that is positioned to the right of another element. For example, if you have a form with a label element on the left and an input field on the right, you can use the rightOf locator to locate the input field by specifying the label element as the reference element:

  ```
  rightOf('label').find('input')
  ```

  Karate will find an input field by default if you omit the .find method:

  ```
  rightOf('label')
  ```

- leftOf: This locator is used to locate an element that is positioned to the left of another element. This works in the same way as the rightOf locator otherwise:

  ```
  leftOf('label').find('input')
  ```

- above: This locator is used to locate an element that is positioned above another element. This can be useful for cases such as looking for a header section above a content section on a page:

  ```
  above('.content').find('.header')
  ```

- `below`: This locator is used to locate an element that is positioned below another element:

  ```
  below('.header').find('.content')
  ```

- `near`: This locator in Karate is used to locate an element that is positioned near another element, regardless of its direction. This is useful when the relative position of the target element is not fixed, and it can appear in different positions around the reference element. This can happen on websites that support **right-to-left languages (RTLs)** such as Arabic.

  For example, this line would find a *button* near the `h1` (first headline) element:

  ```
  near('h1').find('button')
  ```

  It's important to note that the `near` locator may not always return the expected element if there are multiple elements positioned near the reference element. In such cases, it may be necessary to use additional locators or refine the CSS selector to identify the correct target element.

In general, these friendly positional locators are designed to be more resilient to changes in the UI, as they can adapt to modifications such as changes in the layout and style of the web page. Thus, they can be a great alternative to the more fixed locators.

Using wildcard locators

In Karate, wildcard locators use a combination of text and wildcards to identify an element on a web page. The wildcards can match any part of an attribute or text content, making them useful for identifying elements that have dynamic or changing values. You can use these in any place where you can use XPath or CSS locators.

There are different wildcard locators that you can use as shortcuts:

- **By exact text content**: You can search for elements by text using the `{}text` syntax. You can add a tag name in between curly braces, such as `{'h1'}Hello`, or leave them empty, like this: `{}Hello`. The first example will look for the first `h1` element that has the exact text `Hello`, whereas the second will find the first instance of *any* element having this exact text.

- **By containing text content**: If you use a `^` symbol inside curly braces, it will locate the elements by *partial* text. For example, `{'^h1'}Hello` would locate the first `h1` element in which the text contains `Hello`. Using this without a tag—for example, `{'^'}Hello`—would do the same but without a concrete element.

- **By element position**: You can add a position inside curly braces following a `:` instance, like this: `{'h1:2'}Hello`. This would locate the *second* `h1` element that has the exact text `Hello`.

  To find any tag, you can omit the tag name and replace it with an asterisk, like so: `{'^*:2'}Hello`. This would translate to "find the *second element* (regardless of its tag) that contains *Hello* in its text."

It is also possible to use hierarchies between curly braces using XPath instead of simple tag names, but this is needed only in extreme cases.

Determining element locators

The easiest way to find locators to target elements is by using Chrome DevTools within the Chrome browser.

It is generally a good starting point to open the DevTools with the website under test by pressing *F12* in Chrome or going to the *three-dots menu button* in the upper-right corner and selecting **More Tools | Developer Tools**.

Figure 9.1 shows the DevTools alongside an opened website. Clicking on the *Element selection button* (shown as a square with a mouse pointer inside) allows us to target specific elements that we want to examine further:

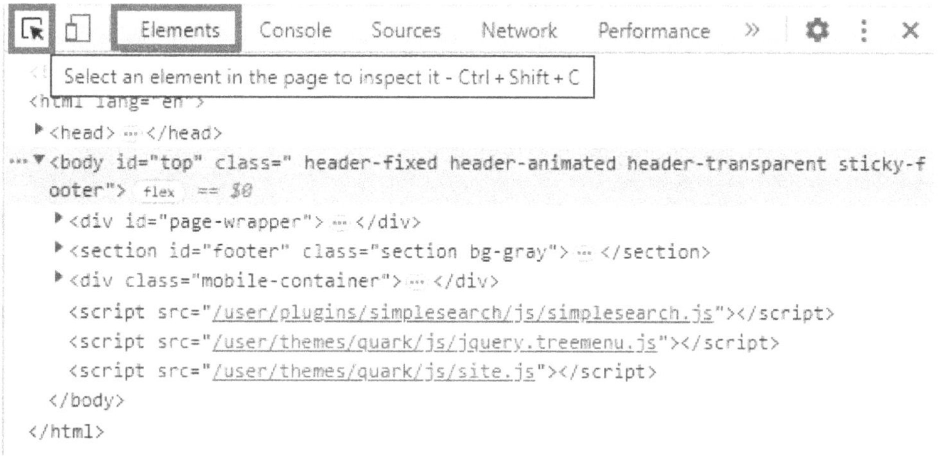

```
 ☐  ☐    Elements    Console    Sources    Network    Performance    »   ⚙  ⋮  ✕

     ⟨  ⟨  Select an element in the page to inspect it - Ctrl + Shift + C
    <html lang="en">
    ▶ <head> ··· </head>
··· ▼ <body id="top" class=" header-fixed header-animated header-transparent sticky-f
      ooter">  flex   == $0
      ▶ <div id="page-wrapper"> ··· </div>
      ▶ <section id="footer" class="section bg-gray"> ··· </section>
      ▶ <div class="mobile-container"> ··· </div>
        <script src="/user/plugins/simplesearch/js/simplesearch.js"></script>
        <script src="/user/themes/quark/js/jquery.treemenu.js"></script>
        <script src="/user/themes/quark/js/site.js"></script>
      </body>
    </html>
```

Figure 9.1 – DevTools element selection

The **Elements** tab that is shown beside it contains the source code of the whole page and highlights the lines inside the HTML code that belong to the chosen element.

Alternatively, you can also activate this view by right-clicking on a web element and choosing **Inspect** from the context menu, as shown here:

Figure 9.2 – Using the Inspect menu

Figure 9.3 shows the DevTools in element selection mode. If you hover over the search input box, it is highlighted and has a tooltip that shows information such as its CSS classes, color, and size. Also, as mentioned, the HTML code belonging to this input element is highlighted on the right:

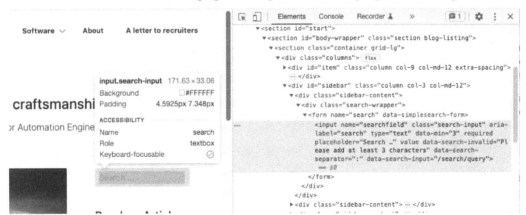

Figure 9.3 – Exploring elements in Chrome DevTools

Based on this information, we can select a good locator to use in our Karate test. It should have these characteristics:

- **Uniqueness**: The locator should identify a single element on the page and not multiple elements unless you want to target a list of elements.

- **Stability**: The locator should be stable and robust and not change frequently as this may lead to test failures.

- **Readability**: The locator should be easy to read and understand so that it can be easily maintained and updated.

- **Consistency**: The locator should ideally be consistent across different browsers and platforms if you want to test multiple browsers and devices using the same test code. Otherwise, you would have to find different locators for the same element based on the different platforms. The worst case would be having to write duplicated scenarios that target specific platforms.

- **Maintainability**: The locator should be easy to maintain and update as the web application evolves. This may be necessary when new features are added or an application is refactored.

Let's look at the source code for the chosen input field again:

```
<input name="searchfield" class="search-input" aria-
label="search" type="text" data-min="3" required=""
placeholder="Search …" value="" data-search-invalid="Please add
at least 3 characters" data-search-separator=":" data-search-
input="/search/query">
```

The highlighted code portions show good characteristics we can potentially use as our locators. For example, we can see that this is the only element having the `search-input` CSS class (we can see this on the left side of *Figure 9.3* where the tooltip shows `input.search-input`).

Let's use this one in the next step!

Interacting with elements

We just determined a locator for the search input box on the website under test. To interact with it, Karate offers a set of specific methods from its driver.

Some of these methods are listed here:

- `click(locator)`: Clicks on the element matching the given locator

- `input(locator, value)`: Sets the value of the input element matching the given locator

- `select(locator, value)`: Selects the option with the given value from the drop-down element matching the given locator

- `scroll(locator)`: Scrolls to the element matching the given locator

- `back()`: Navigates the browser back to the previous page

- `forward()`: Navigates the browser forward to the next page

- `refresh()`: Refreshes the current page

- `close()`: Closes the current browser window

We can now put the information we retrieved from the DevTools together to create our first scenario.

Using interactions in a test scenario

So far, we are just opening the website under test but there is no interaction yet. So, we need to create some additional code, as follows:

```
Scenario: First test
    * configure driver = { type: 'chrome' }
    Given driver 'https://softwaretester.blog'
    When input('.search-input', ['Magic', Key.ENTER])
    And waitForUrl('search/query:Magic')
```

Here, I added two steps:

1. As we have seen before, we can target the search input box on the website with the .search-input locator. Now, we can fill it with our Magic search term using Karate's input function. Note that I use a special syntax to first write the Magic string into the text field and finish by pressing the *Enter* key (specified by Key.ENTER). This is done by using an array of inputs as the second parameter instead of a single string.

2. The waitForUrl function is very handy if an action on a website navigates to a different URL. This is the case here: searching for Magic will go to the https://softwaretester. blog/search/query:Magic URL. However, adding the complete URL is not necessary here as the waitForUrl function waits for a URL containing the specified string. So, search/ query:Magic is enough in this case.

If we run this test now, it should pass and produce the following Karate test report:

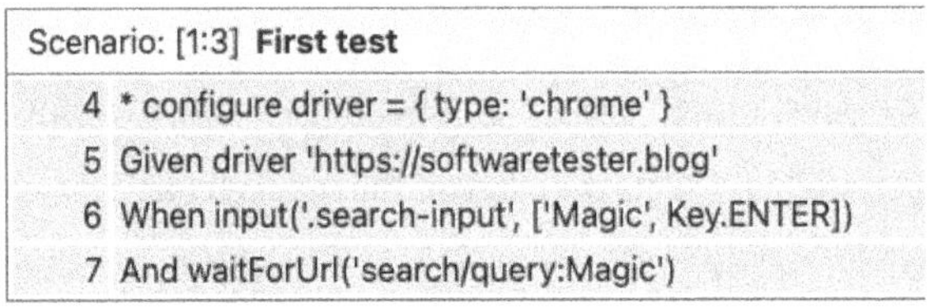

Scenario: [1:3] **First test**

4	* configure driver = { type: 'chrome' }
5	Given driver 'https://softwaretester.blog'
6	When input('.search-input', ['Magic', Key.ENTER])
7	And waitForUrl('search/query:Magic')

Figure 9.4 – Successful UI test

Chrome error

For some Chrome versions (for example, *111*) you might get an error saying the following:

```
com.intuit.karate - driver config / start failed: io.netty.
handler.codec.http.websocketx.WebSocketClientHandshakeException:
Invalid handshake response getStatus: 403 Forbidden, options:
{type=chrome, target=null}
```

If this happens, you can replace the first step of this scenario with `* configure driver = { type: 'chrome',addOptions: ["--remote-allow-origins=*"] }`.

To see the negative case, let's change the `waitForUrl` step condition to wait for `search/query:Wrong` for fun and run the test again. In this case, we get the expected failure in the logs and test report:

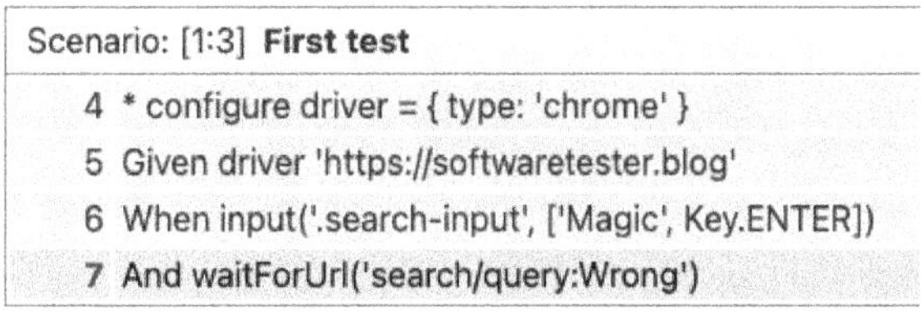

Scenario: [1:3] **First test**

4 * configure driver = { type: 'chrome' }
5 Given driver 'https://softwaretester.blog'
6 When input('.search-input', ['Magic', Key.ENTER])
7 And waitForUrl('search/query:Wrong')

Figure 9.5 – Failed UI test

Let's look at two other browser test examples to showcase some important additional functionality.

Making web element assertions

When testing in the browser, a big part of these tests is not only element interactions but also waiting for conditions and matching data. This next scenario uses a website called `https://www.demoblaze.com`, which is a special public site to practice browser automation with. It simulates a webshop that offers phones, laptops, and monitors.

As before, we will start by opening the website:

```
Scenario: Check product name
   * configure driver = { type: 'chrome' }
   Given driver 'https://www.demoblaze.com'
   And driver.dimensions = {x: 0, y: 0, width: 1024, height: 768
}
```

In this case, we also set the browser window size and position using the `driver.dimensions` configuration. This way, it is possible to test specific viewports quickly and easily. The {x: 0, y: 0, width: 1024, height: 768 } value positions the browser at the 0, 0 screen coordinates (the top-left corner) and sets it to 1,024 by 768 pixels, as shown in the following screenshot:

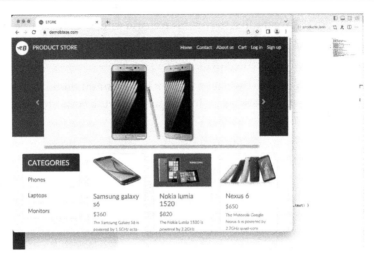

Figure 9.6 – Demoblaze website in a resized browser

The goal of the test is to navigate to the **Monitors** category and verify that the product name of the first shown product is **Apple monitor 24**, as is the case here:

Figure 9.7 – Headline to be verified in the test

We will first use a wildcard locator to click on the **Monitors** category by targeting `{a}Monitors`, meaning the first link whose text is exactly `Monitors`:

```
And waitForEnabled('{a}Monitors').click()
* waitForUrl('/#')
```

Note that I use the waitForEnabled method here to make sure that this link is visible and ready for interaction before clicking it. This may be necessary depending on how the website is developed. Also, note that we can directly chain the call to the click method without adding a separate step for it.

After clicking, waitForUrl('/#') pauses until the URL contains /#. This happens when clicking any category navigation link. This is typically something you can find out by exploring the website first before starting to write a test for it.

Now, we need to get the title of the first product. This is again a case for the DevTools:

Figure 9.8 – Checking the locator of the targeted element

Fortunately, the titles of the listed products all have the card-title class, so we can use the following code to get the text of the first element having this class:

```
* def productName = text('.card-title')
* match productName == 'Apple monitor 24'
```

As with API tests, we can define variables to store values in. In this case, the element text is stored in the productName variable that we can finally match to our expected Apple monitor 24 string.

Let's look at another example that uses element lists instead of a single web element.

Making web element list assertions

This next example uses yet another demo website to practice testing with: https://computer-database.gatling.io. This is the frontend to a computer database where we will test the search functionality:

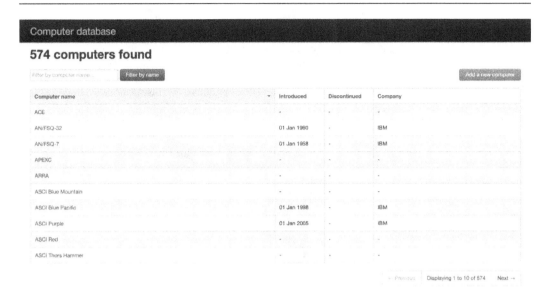

Figure 9.9 – The computer database website

We start with the base of the scenario again—configuring the driver and opening the website:

```
Scenario: Working with element lists
  * configure driver = { type: 'chrome' }
  Given driver 'https://computer-database.gatling.io/'
  * driver.maximize()
  * def searchTerm = 'MacBook'
```

Additionally, we maximize the browser using the `driver.maximize` function. Also, we create a new `searchTerm` variable. This is convenient because the goal of this test is to search for `MacBook` and ensure that all found entries include this term.

Through the help of the DevTools, we can see that the search input field has an ID of `searchbox` and the search button's ID is `searchsubmit`:

```
▼<form action="/computers" method="GET">
    <input type="search" id="searchbox" name="f" value placeholder="Filter by computer nam
    e..." required="required"> == $0
    <input type="submit" id="searchsubmit" value="Filter by name" class="btn primary">
    <a class="btn success" id="add" href="/computers/new">Add a new computer</a>
  </form>
```

Figure 9.10 – Locating the IDs of web elements

Now, we can continue our test scenario with these steps:

```
When input('#searchbox', searchTerm)
And click('#searchsubmit')
```

This enters our MacBook search term into the search field and clicks the **Search** button afterward:

10 computers found

| MacBook | Filter by name | Add a new computer |

Computer name ▾	Introduced	Discontinued	Company
MacBook	16 May 2006	-	Apple Inc.
MacBook 13-inch Core 2 Duo 2.13GHz (MC240LL/A) DDR2 Model	-	-	Apple Inc.

Figure 9.11 – Result list on the website

This is how the result list looks on the website. When we explore the HTML, we can clearly see that the parent element is `table`. Each of its result rows is a `tr` element inside `tbody`. The first `td` element in each row contains the computer name that we are looking for:

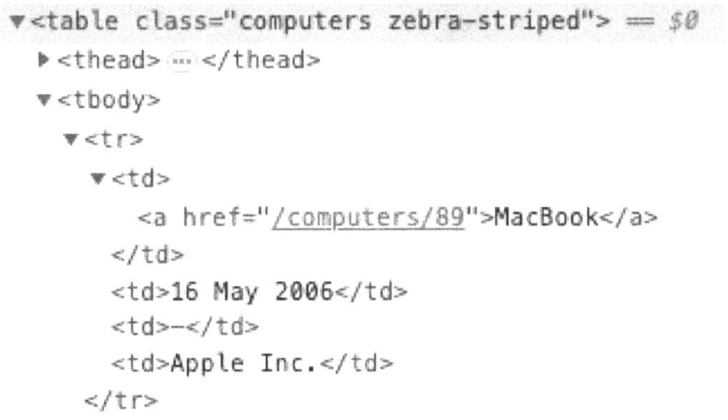

```
▼<table class="computers zebra-striped"> == $0
   ▶<thead> ⋯ </thead>
   ▼<tbody>
     ▼<tr>
       ▼<td>
           <a href="/computers/89">MacBook</a>
         </td>
         <td>16 May 2006</td>
         <td>-</td>
         <td>Apple Inc.</td>
       </tr>
```

Figure 9.12 – HTML code of the result list table

With this information, we can now retrieve all computer names to compare them to the search term. For this, we use a mechanism that we have explored in API tests with Karate before:

```
* def resultRows = locateAll('table.computers tbody tr')
* assert karate.sizeOf(resultRows) > 0
```

The `locateAll` function takes a locator and returns *all* matching elements, not only the first match. So, unlike before, we get a list of all rows in the table with our search results. For this, we use the `table.computers tbody tr` CSS locator, which translates to: return all `tr` elements within `tbody` that are contained within a `table` that has the `computers` class. This result is then saved in the `resultRows` variable. Also, we check if there are more than zero results using `karate.sizeOf`.

Now, we can finish the scenario using these steps:

```
* def computers = []
* def getNames = function(row) { karate.appendTo(computers,
row.children[0].text) }
* karate.forEach(resultRows, getNames)
* match each computers contains searchTerm
```

Let's walk through what happens here:

1. We create a new empty array called `computers` that should contain all computer names in the end.

2. Also, we define a new function called `getNames`. This takes a result row web element and retrieves the text of its first element. If you recall, the very first child of each `tr` element is the `td` element containing the computer name. This is also shown in *Figure 9.12*.

    The `row.children[0].text` code lets us target this first child of a row and return all text contained within it. Also, it adds the retrieved text to our `computers` array using `karate.appendTo`.

3. By writing `karate.forEach(resultRows, getNames)`, we go through every result row and call the `getNames` function on it.

4. In the end, our `computers` array contains all computer names, and we can match each one to see if it contains our `searchTerm` variable with the `MacBook` value we defined before.

Now that we have some running Karate UI tests, we will look at some powerful debugging features that come with Karate UI and the VS Code extension.

Debugging UI tests

In *Chapter 4*, *Running Karate Tests*, we saw Karate's powerful debugging features in combination with VS Code. We will now see that we can also use the same techniques when dealing with browser tests.

Note that these are techniques and approaches that can be used in all supported browsers. If you want to know more options that are specific to the target browser and driver in use—for example, video recording with the Karate Docker image—please refer to this section of the official documentation: `https://github.com/karatelabs/karate/tree/master/karate-core#configure-drivertarget`.

Let's now explore which debugging options Karate offers to simplify and speed up browser test development.

Taking screenshots

Karate will automatically take a screenshot if a test fails. This screenshot is automatically attached to the test report and can be seen by clicking on the failed step of the test. This makes it easier to debug the cause of the failure:

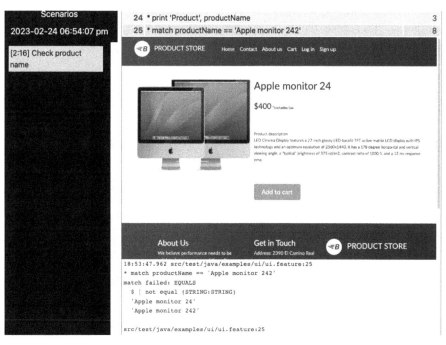

Figure 9.13 – Screenshot of a failed test in Karate's test report

Apart from this default functionality, we can also add screenshots whenever we want. This can be a nice tool to help follow what the test is doing.

Taking a screenshot of the current state of the page is a matter of adding this step whenever an image should be taken:

```
* screenshot()
```

This screenshot will then be attached at this exact point in the test report.

> **Visual testing**
>
> Karate UI supports **visual testing**, meaning taking a screenshot of a web page and comparing it to a known baseline image that shows the desired state. It can then report the difference percentage and highlight changes.
>
> This capability requires a more elaborate setup and custom code to include, so it is out of scope for this chapter.
>
> If you want to know more about it, please check the official documentation here: `https://karatelabs.github.io/karate/#compare-image`.

The beauty of this screenshot capability is that we can also choose any element on the web page to take a screenshot of. This reduces the size of the screenshot and removes clutter when the focus is to be directed to a specific element. The code is shown here:

```
* screenshot('locator')
```

Here, `locator` needs to be replaced with the locator of a web element.

Highlighting elements

For visual confirmation of elements, you can highlight a single element or a group of similar elements. These are then marked clearly with a border and background color while the test is running.

The `highlight` function marks the first found element matching a locator, while the `highlightAll` function does the same for all found elements.

You can see an example output here:

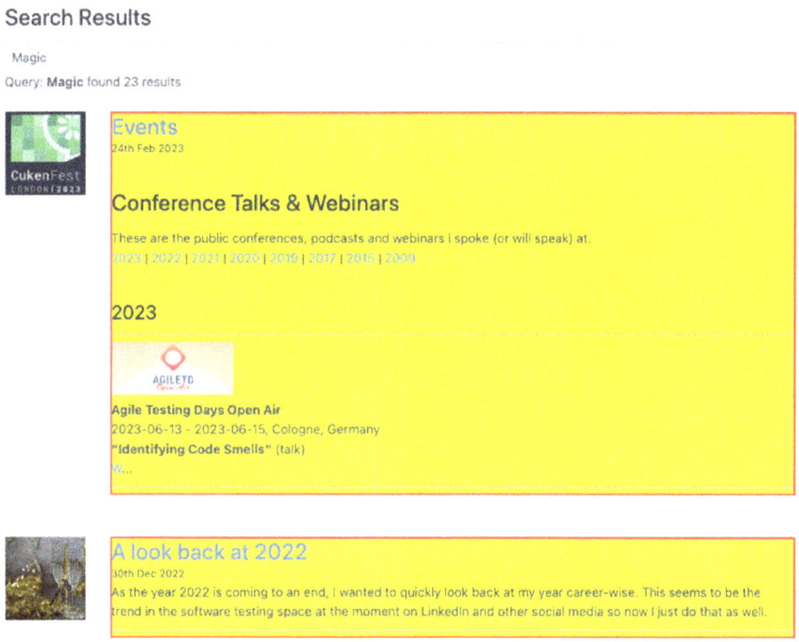

Search Results

Magic

Query: **Magic** found 23 results

Events

24th Feb 2023

Conference Talks & Webinars

These are the public conferences, podcasts and webinars I spoke (or will speak) at.

2023 | 2022 | 2021 | 2020 | 2019 | 2017 | 2016 | 2009

2023

AGILETD

Agile Testing Days Open Air

2023-06-13 – 2023-06-15, Cologne, Germany

"Identifying Code Smells" (talk)

W...

A look back at 2022

30th Dec 2022

As the year 2022 is coming to an end, I wanted to quickly look back at my year career-wise. This seems to be the trend in the software testing space at the moment on LinkedIn and other social media so now I just do that as well.

Figure 9.14 – Highlighted elements during the test run

In the preceding screenshot, I used the following step to highlight all elements with the `search-item` class, which is the common class in my website that is shared by each search result:

```
* highlightAll('.search-item')
```

The default highlighting duration is 3 seconds. As always, you can configure this within a feature file using Karate's `configure` keyword followed by `highlightDuration` or do the same in `karate-config.js` using the `karate.configure` function.

Exporting PDFs

Sometimes, it can be handy to take a complete image of the current state of the website and save it as a PDF. For example, this could be helpful when adding this to a bug ticket to provide visual clues to developers.

This is as simple as calling these steps:

```
* def resultAsPdf = pdf({})
* karate.write(resultAsPdf, "search.pdf")
```

Here, we use the `pdf` function that returns the complete bytecode, which we can then pass to the `karate.write` method to write to a file. It expects a JSON object as a parameter for configuration (you can pass `{'orientation': 'landscape'}` to force it into landscape mode). In this example, though, I used an empty object (`{ }`) to use the default configuration:

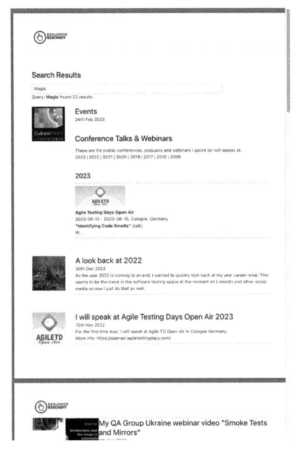

Figure 9.15 – Generated full-page PDF

This PDF will keep the website links as well so that you can directly jump to the real website by clicking on these!

The generated PDF is automatically stored under the specified name within the `target` directory of our test project.

Using the debugger

We used the Karate debugger for API tests in *Chapter 4, Running Karate Tests*. As described there, the prerequisite to use the debugger is to add a debug configuration to the launch.json file in VS Code.

As a recap, the quickest way to do this is to click on **Karate: Debug** in VS Code's CodeLens, which appears above test scenarios. This opens a dialog where you can select **Karate (debug)**, which then brings you to the launch.json file automatically:

Figure 9.16 – Activating Karate (debug)

With the **Add Configuration…** button, you open a list of configurations. Select **Karate (debug):
Maven** and save the file:

```
{
    // Use IntelliSense to learn about possible attributes.
    // Hover to view descriptions of existing attributes.
    // For more information, visit: https://go.microsoft.com/fwlink/?linkid=830387
    "version": "0.2.0",
    "configurations": [
        |
        {} Jest: create-react-app (ejected)
    ]   {} Jest: Default jest configuration
}       {} Karate (debug): Gradle
        {} Karate (debug): Maven                        Add Configuration...
        {} Karate (debug): NPM
        {} Node.js: Attach
EMS  2  {} Node.js: Attach to Process
        {} Node.js: Attach to Remote Program            Karate Runner - Task  ✓  + ∨ [
:/java/exa {} Node.js: Electron Main
        {} Node.js: Gulp task
Scanning {} Node.js: Launch Program
-------- {} Node.js: Launch via npm
Building
```

Figure 9.17 – Adding a Maven debug configuration for VS Code CodeLens

The next time you press the **Karate: Debug** link in CodeLens, it should launch the debug server and put VS Code into debug mode.

Now, you can debug browser tests the same way as you can with API tests. You can refer to *Chapter 4* again to see how to execute steps one by one, jump back and forth between tests, and even change test code on the fly.

When opening the Debug console when the test execution is halted at a breakpoint, you can execute all the commands that you can use within Karate tests. This lets you try out locators to find elements or interact with them directly, which can speed up test development drastically:

Figure 9.18 – Debug console command-line usage

In this case, I typed `highlightAll('a')` into the command line at the bottom of the Debug console to frame all link elements of the website:

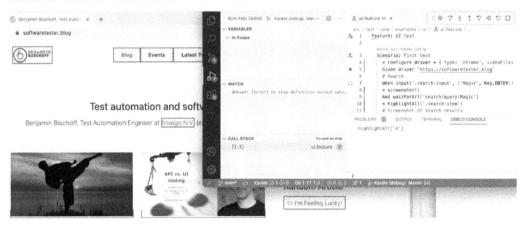

Figure 9.19 – Result of the highlightAll('a') command in the Debug console

Pressing *Enter* will execute this command immediately and perform the requested action! The nice thing about debugging in Karate UI is that you can immediately confirm these actions visually in the opened browser.

Using karate.stop to pause execution

The `karate.stop` method is a special way to debug that also works outside of debug mode. It pauses the test execution and lets you take over the browser to check elements using Chrome DevTools or confirm that actions that normally execute very quickly are happening correctly.

You can add this at any point in your test scenario using this step:

```
* karate.stop(1234)
```

The number *1234* can be freely chosen. It represents a port number. When running a test containing such a `karate.stop` step, execution halts and the console log contains a line like this:

```
*** waiting for socket, type the command below:
curl http://localhost:1234
in a new terminal (or open the URL in a web-browser) to proceed
...
18:51:50.119 [stop-listener-1234] INFO  c.i.karate.shell.
StopListenerThread - starting thread: stop-listener-1234
```

You can see that Karate internally starts a special thread that waits for a signal on the port you specified before. To continue the execution of the test, you can either open the shown URL in a browser or execute the `curl` command with the shown URL in a terminal window.

Please be aware that, even if you close the browser, Karate will not automatically stop execution when the waiting thread is active. You must actively do one of the described steps to do so. Also, note that when opening the URL in a browser, you will probably get an error page stating ERR_CONNECTION_REFUSED. This is expected as this URL is just a trigger for Karate and not a real website URL.

It is very important to mention that this functionality is only intended for the development or debugging of tests, and it should never end up in a production test suite.

Karate's power lies in all its different modules working seamlessly together. In the next section, we will use it in combination with Karate's mocking feature we explored before.

Redirecting HTTP calls

If you remember, in *Chapter 6, More Advanced Karate Features*, we had an example of authenticating an API using an authentication token. Now, imagine—for example—having a real login to a website performed in Karate UI and then reusing the authentication token in Karate API requests. In another example, you could get the order number for a certain product from a store API using Karate API requests and then automatically enter it into the search mask on a website to further test the UI with this exact product.

Here, it becomes clear how powerful the combination of these two Karate test modules is. However, the whole situation becomes even more powerful when we use the mocking capabilities that we already learned about in *Chapter 7, Customizing and Optimizing Karate Tests*!

> **Chrome only**
>
> This mocking capability is *only available in Chrome* when using the Karate *Chrome* driver (the *ChromeDriver* driver will not work since it uses the WebDriver protocol, which does not support the interception of network requests).
>
> This, again, makes it apparent that generally, Chrome is a great starting point to test the most important features since it offers the most functionality through its DevTools protocol integration.

In the example in the next section, we will use a Karate test double to intercept specific calls of the web browser and use our own mock data to make tests quicker and more predictable.

Investigating API requests

The first step is to find out which endpoints a website communicates with to retrieve data. We will use the `https://www.demoblaze.com` site again since it probably receives its products from a REST endpoint. Let's try to find out where the data for these product descriptions comes from!

Figure 9.20 – The website under test

To check out where data comes from, we can use the DevTools again. This time, we will open them before accessing the website. Also, we need to switch to the **Network** tab, which will show us the website's communication with external data sources:

Figure 9.21 – Network inspection in the DevTools

To only see the data sources, click on **Fetch/XHR**. This filters out all other sources, such as image requests and JavaScript that we don't need to see. In this case, there is one data source called `entries` that is interesting to explore further:

Figure 9.22 – Determining the request URL

A click on **Headers** shows us exactly which endpoint was hit by the website. To see the data, we can switch to **Preview**, which presents the complete JSON data that was retrieved!

Figure 9.23 – Copying a JSON response from the DevTools

Let's copy this JSON structure for reference using the right-click menu and selecting **Copy object**. We will use this data in the next step to create our mock-up.

Adding a mock response

We will take the real response we copied before as a basis to create our mock. This way, we can start with the real data structure first and then gradually explore what happens if we change values or add and remove data.

This is a portion of the copied response from the DevTools:

```
{
  "Items": [
    {
      "cat": "phone",
      "desc": "The Samsung Galaxy S6 is powered by 1.5GHz octa-
core Samsung Exynos 7420\n processor and it comes with 3GB
of RAM. The phone packs 32GB of \ninternal storage cannot be
expanded. ",
      "id": 1,
      "img": "imgs/galaxy_s6.jpg",
      "price": 360,
      "title": "Samsung galaxy s6"
    },
    ...
  ],
  "LastEvaluatedKey": {
```

```
      "id": "9"
    }
  }
```

Let's now create a file called `products.json` that should be used as a replacement for the real API response. For this purpose, I removed all except one item and changed the `desc`, `price`, and `title` values.

For the `img` value, I used the `https://dummyimage.com` website, which provides custom placeholder pictures, mainly for creating mock-up designs for websites. By using the `https://dummyimage.com/200x200?text=Karate` URL, it will return a 200 x 200-pixel image containing the `Karate` text.

Also, I assumed that we need the `LastEvaluatedKey` value, so I left it there.

This approach of not changing too much at once is often a good idea when dealing with API responses. This way, you don't end up introducing errors that are hard to backtrack:

```
{
    "Items": [
      {
        "cat": "phone",
        "desc": "This is a test product",
        "id": 1,
        "img": "https://dummyimage.com/200x200?text=Karate",
        "price": 123,
        "title": "Test1"
      }
    ],
    "LastEvaluatedKey": {"id": "9"}
}
```

This is the complete mocked response that should be served in place of the real API call in the next step.

Configuring the mock

We already know from the network request exploration that the website calls the `entries` endpoint to retrieve the product data. So, we can now use Karate's mocking ability, which we used in *Chapter 7, Customizing and Optimizing Karate Tests*.

The first thing is to create a feature that serves our mock:

```
@ignore
```

```
Feature:

  Background:
    * configure cors = true

  Scenario: pathMatches('/entries')
    * def response = read('products.json')
```

Here, we first set the special cors variable to true, which sets some headers that enable us to serve our own data in place of the original REST call without security warnings. This must be done in the Background section; otherwise, it does not have any effect.

> **CORS**
>
> **Cross-Origin Resource Sharing (CORS)** is a web browser feature that is related to security. It restricts web pages from making requests to a different domain than their own. This feature exists to keep malicious websites from having access to sensitive information from any other domains. For our tests, we must actively allow this access so that the live website can retrieve data from our local mock server.

With pathMatches('/entries'), we can tell this mock that it should react to calls to any path that contains /entries. Finally, response = read('products.json') takes our mock JSON file and serves this as the response for this endpoint.

So, as the last step, we must redirect a real request to our mock.

Using the mock in a test

Let's now see how we can use this in a test scenario:

```
Feature: Intercept test

  Scenario: Inject custom products
    * configure driver = { type: 'chrome' }
    * driver 'https://www.demoblaze.com'
    * driver.intercept({ patterns: [{ urlPattern: '*/entries'
}], mock: 'products.feature' })
    * karate.stop(5555)
```

The key here is the driver.intercept method. This can be used to forward specific requests to a mock feature. It accepts two parameters: patterns, to set the URL pattern it should look for (in this case, */entries), and mock, which points to our mocked products.feature.

I ended this feature with the already-discussed `karate.stop` command to keep the browser open so that we can visually verify that our mock was used. It makes sense to remove this step later and replace it with specific assertions or matchers to ensure that the correct data is displayed in the right web elements. This is the result of our work:

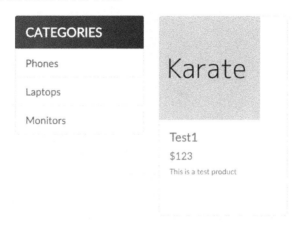

Figure 9.24 – Injected mock data

The website still makes requests to the real endpoint but our test intercepts this call and simulates the API by returning our mocked data right away. This is a very powerful feature of the CDP and Karate UI combination that can be a great solution for a lot of use cases.

Summary

In this chapter, we explored an area of Karate that deviates from its core API testing functionality. Even though API calls are used in the background, browser tests feel very different in that they focus on the UI of a web application.

First, we saw what browser tests are used for, why they matter, and that there are a lot of competing tools in this area between which Karate stands out due to its API focus. Specifically, we looked at how to write basic tests in Chrome and why this should be the first browser to start with.

Following this, we learned how to find and interact with web elements and saw what possibilities there are to speed up test development using Karate's debugging functions.

Finally, we combined Karate's mocking abilities with API call interception to bring this topic full circle.

In *Chapter 10, Performance Testing with Karate Gatling*, we will look at the last specialty of Karate: running performance and load tests against APIs to determine their resilience.

10

Performance Testing with Karate Gatling

So far, we have seen a lot of different aspects of Karate API testing, including browser testing. In this chapter, we will explore yet another kind of testing that Karate offers – performance testing. We will use a simple Karate scenario and walk through the necessary steps to run it as a performance test. Also, we will look at logs and test reports to understand more about these types of testing and the conclusions we can draw from these.

In this chapter, we will cover these main topics:

- What is Gatling?
- Setting up Karate Gatling
- Running Karate tests as performance tests
- Checking out Gatling reports

Technical requirements

The code examples for this chapter can be found in this book's GitHub repository: `https://github.com/PacktPublishing/Writing-API-Tests-with-Karate/tree/main/chapter10`.

You will require the following:

- The system and IDE setup we completed in *Chapter 2*, *Setting Up Your Karate Project*

Creating the test scenario

For the test scenario of this chapter, we will keep it simple. We will use the so-called *Jikan API* (`https://jikan.moe`), which provides information about anime and manga movies, series, and books. As this demonstration should show some statistics about general API availability, we will not bother with its detailed request and response data for now and instead concentrate only on the response code of an individual endpoint.

This is the test scenario we will use:

```
Feature: Performance testing

  Scenario: Performance
    * url 'https://api.jikan.moe/v4/'
    * path 'anime/1'
    When method get
    Then status 200
```

At this point in the book, this test scenario should be self-explanatory, as it does not include any fancy logic. The only thing that is done here is to perform a GET request on the anime/1 path of this API (which provides a lot of information about an anime called *Cowboy Bebop*). As mentioned, the only assertion that is made here is to check for the 200 (OK) HTTP status code.

We can run it now from the Terminal by calling mvn clean test, like all the other Karate tests we have created before. This will be the basic test we use to explore Karate Gatling.

What is Gatling?

Load and performance testing are two types of software testing that help developers to evaluate the performance of their applications under different workloads. Load testing is the process of simulating user traffic on an application to test its ability to handle a high volume of requests. This type of testing helps to identify the maximum capacity of an application and determine whether it can handle the expected workload without crashing or slowing down.

On the other hand, performance testing is the process of measuring the speed, scalability, and stability of an application under different conditions. This type of testing helps to identify the performance bottlenecks in the application and optimize it for better response and resilience.

In the context of APIs, load and performance testing are used to evaluate the ability of an API to handle a high volume of requests and respond within the expected time frame. These tests help developers to identify potential issues such as slow response times, errors, unexpected behavior, or even complete crashes that can impact the user experience negatively.

There are multiple tools in software development for this purpose. Examples are **Apache JMeter** (https://jmeter.apache.org), **Locust** (https://locust.io), or **Siege** (https://www.joedog.org/siege-home). This chapter will be about another common tool that has great strength and works well with Karate – Gatling.

Gatling (`https://gatling.io`) is an open source load and performance testing tool that helps developers to simulate and measure the performance of their applications. It uses the **Scala** programming language and is designed to be efficient, scalable, and easy to use.

> **More about Scala**
>
> Scala is a general-purpose programming language that was designed to be highly scalable and efficient for large-scale applications. It combines functional programming with object-oriented programming, allowing developers to write concise and expressive code that is easy to read and maintain. It has gained popularity in recent years because it is compatible with Java and supports distributed computing. This makes it a popular choice for creating high-performance applications.
>
> More information on Scala can be found here: `https://www.scala-lang.org`.

With Gatling, you can create complex load-testing scenarios and execute them on multiple machines to test an application or APIs under different conditions. This is typically done by installing Gatling and using its user interface to design test cases, which can then be run in multiple threads to simulate multiple requests at once.

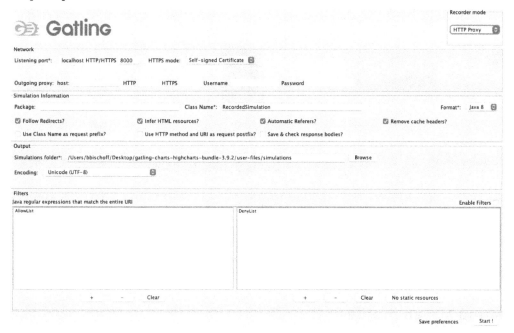

Figure 10.1 – The Gatling user interface

In this chapter, we will not install Gatling or use the desktop application. Instead, our goal is to take one of our already existing Karate tests and run it as a load test using Gatling programmatically.

Gatling provides real-time reporting and analysis of test results in graphical format, which helps to identify performance bottlenecks. This is important information when it comes to optimizing applications to handle a higher load. Gatling is widely used in the software development industry to ensure the quality and scalability of web applications. Later, we will see how to interpret these reports.

Let us now continue by creating a Karate project and going through the necessary steps to integrate Gatling.

Setting up Karate Gatling

We will now set up everything to create our Gatling performance test. Most of the work will be done in the Maven POM file, as you will see later in this section.

Setting up Scala in VS Code

Writing effective Scala code can be tricky. So, it is advisable that our IDE fully supports it. Luckily, there is an official **Scala Syntax** plugin available for VS Code. If you just search for `scala` in VS Code's **Extensions** tab, it should be the first one shown.

Click on **Install** to add it to your IDE.

Figure 10.2 – The VS Code Scala Syntax plugin

This plugin mainly provides code highlighting, which helps us get familiar with this language. Scala is not too different from Java in terms of syntax, though. Also, we will only need one simple Scala class for our test project.

Creating Maven profiles

So far, we have run tests using the `mvn clean test` command, using Maven Surefire. This runs the tests as normal unit tests. To run the tests using Gatling, we could now change the dependencies and the build flow in our `pom.xml` file. This has a major downside – we could not run the same tests as unit tests anymore. Our goal should be to have the choice of whether we want to execute the Karate scenario as a unit test or run it as a performance test. This is where **Maven profiles** come in.

They allow us to organize different dependencies and workflows in one pom.xml file and activate the ones we need at runtime. This makes it possible to choose exactly what tests we want to run and how!

> **More about Maven profiles**
>
> In Maven, a profile is a set of configuration settings that can be used to customize a Maven build for different environments, such as development, testing, or production. Profiles are defined in the project's pom.xml file and can be activated based on a variety of conditions, such as the operating system, a specific Maven property, or the presence of a file. When a profile is activated, its configuration settings override the default settings in the pom.xml file. That means that everything is self-contained and does not require constant file changes. Profiles can be used to configure a variety of build-related settings, such as compiler settings, dependencies, and plugins.

Let's start by moving the blocks in the pom.xml file around to reach our goal of two profiles.

Reorganizing pom.xml

The basis of our work is the standard pom.xml file that is generated by the Karate Maven archetype. This has everything in it to run Karate tests as unit tests as we have seen numerous times before. We will leave the complete beginning of this file alone but reorganize the <build> section a little.

This is the standard build section, which defines the standard test resources that should be considered, the Java version, and the Surefire plugin, which executes test scenarios as unit tests:

```
<build>
  <testResources>
    <testResource>
      <directory>src/test/java</directory>
      <excludes>
        <exclude>**/*.java</exclude>
      </excludes>
    </testResource>
  </testResources>
  <plugins>
    <plugin>
      <groupId>org.apache.maven.plugins</groupId>
        <artifactId>maven-compiler-plugin</artifactId>
        <version>${maven.compiler.version}</version>
      <configuration>
        <encoding>UTF-8</encoding>
```

```
        <source>${java.version}</source>
        <target>${java.version}</target>
        <compilerArgument>-Werror</compilerArgument>
      </configuration>
    </plugin>
    <plugin>
      <groupId>org.apache.maven.plugins</groupId>
      <artifactId>maven-surefire-plugin</artifactId>
      <version>${maven.surefire.version}</version>
      <configuration>
        <argLine>-Dfile.encoding=UTF-8</argLine>
      </configuration>
    </plugin>
  </plugins>
</build>
```

We will only keep the `<testResources>` block inside of `<build>`, as this is required for the unit tests as well as the performance tests. Also, `maven-compiler-plugin` including its configuration will stay, as it defines the Java version our tests should run with.

The highlighted `maven-surefire-plugin` block will be moved to a new profile because this one is only needed to run unit tests.

Adding the default profile

We will now add a new `<profiles>` section to the `pom.xml` file, which includes our two required profile definitions:

- The first is a profile called `default` to run unit tests. This one should be used automatically if no other profile is chosen.

- The second is a profile called `performance` to run Gatling performance tests. This one should be used on demand.

This is the *default* profile that includes `maven-surefire-plugin` from before:

```
<profiles>
  <profile>
    <id>default</id>
    <activation>
      <activeByDefault>true</activeByDefault>
```

```
      </activation>
      <build>
        <plugins>
          <plugin>
            <groupId>
              org.apache.maven.plugins
            </groupId>
            <artifactId>maven-surefire-plugin</artifactId>
            <version>${maven.surefire.version}</version>
            <configuration>
              <argLine>-Dfile.encoding=UTF-8</argLine>
            </configuration>
          </plugin>
        </plugins>
      </build>
    </profile>
  </profiles>
```

Our profile has an id value of default, by which it can be activated via the command line. Since we want this to be active if no other profile is chosen, we can add an <activation> block and include <activeByDefault>true</activeByDefault>.

This *default* profile contains a <build> section of its own, which includes maven-surefire-plugin.

So now, if we run mvn clean test from the command line, it should behave exactly as before these changes. The default profile is activated, and Surefire executes our Karate test scenario.

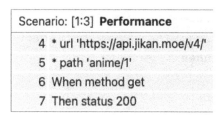

Figure 10.3 – Successful unit test run in the Karate report

This is also reflected in the Karate report as usual. Now, we need an additional profile for Gatling.

Adding the performance test profile

Within the <profiles> section of our pom.xml file, we will now add the performance profile:

```xml
<profile>
  <id>performance</id>

  <dependencies>
    <dependency>
      <groupId>com.intuit.karate</groupId>
      <artifactId>karate-gatling</artifactId>
      <version>${karate.version}</version>
      <scope>test</scope>
    </dependency>
  </dependencies>

  <build>
    <plugins>
      <plugin>
        <groupId>net.alchim31.maven</groupId>
        <artifactId>scala-maven-plugin</artifactId>
        <version>4.8.1</version>
        <executions>
          <execution>
            <goals>
              <goal>testCompile</goal>
            </goals>
            <configuration>
              <excludes>
                <exclude>java</exclude>
                <exclude>**/*.java</exclude>
              </excludes>
            </configuration>
          </execution>
        </executions>
      </plugin>
      <plugin>
        <groupId>io.gatling</groupId>
        <artifactId>gatling-maven-plugin</artifactId>
        <version>4.3.0</version>
```

```
        <executions>
          <execution>
            <phase>test</phase>
            <goals>
              <goal>test</goal>
            </goals>
          </execution>
        </executions>
      </plugin>
    </plugins>
  </build>
</profile>
```

It has the `performance` ID, so it is possible to choose this later on the command line. It does not have an `<activation>` block defined, so it is completely ignored in the default run.

Let's look at each dependency and plugin within this profile:

1. **karate-gatling**: This is the library that enables a bridge between Karate and Gatling. Conveniently, its version is kept in sync with the `karate-junit5` dependency, so we can use the same version definition here.

2. **scala-maven-plugin**: This plugin is important to compile the Scala code we will write in the next section. Since we are using it for tests, we need to specify the `testCompile` goal to make sure that the Scala class is compiled and ready before the test phase begins. Here, we can add an `<excludes>` block that prevents `scala-maven-plugin` from compiling the Java classes in our test project, since it should only take care of the Scala code. More information about this plugin can be found at `https://davidb.github.io/scala-maven-plugin`.

3. **gatling-maven-plugin**: This plugin lets you run Gatling tests on the command line and does not require you to have the Gatling standalone installation on your system. We want to run them within the normal Maven test phase, so we can run these using the `mvn clean test` command. That's why this block is required here:

```
<executions>
  <execution>
    <phase>test</phase>
    <goals>
      <goal>test</goal>
    </goals>
  </execution>
</executions>
```

More information about this plugin can be found at `https://gatling.io/docs/gatling/reference/current/extensions/maven_plugin`.

We have our complete Maven setup now with the two profiles. We already tested the *default* one, and we will soon be ready to use the `performance` profile as well.

The next step is to create a Gatling simulation that can run our Karate scenario with the number of virtual users and the specific time span we want to test.

Running Karate tests as performance tests

Now that we finally have the complete Maven setup and our basic test scenario, we can work on running this as a performance test.

Creating a simulation

We need to create a new Scala file that tells Gatling that it should run our new test scenario. Also, we must specify which values we want to use for our number of virtual users and how long we want our performance test to run.

We will call the `MySimulation.scala` file and place it in the same folder as the feature file. You can, of course, choose to move it somewhere else, but in this case, I prefer to have the `MySimulation.scala` file and the feature file together:

```scala
package examples.performance

import com.intuit.karate.gatling.PreDef._
import io.gatling.core.Predef._
import scala.concurrent.duration._

class MySimulation extends Simulation {

  val protocol = karateProtocol()
  protocol.nameResolver = (req, ctx) =>
    req.getHeader("karate-name")
  protocol.runner.karateEnv("perf")

  val myScenario = scenario("My great Scenario")
    .exec(karateFeature(
      "classpath:examples/performance/performance.feature")
    )
```

```
setUp(
  myScenario.inject(rampUsers(10).during(10))
    .protocols(protocol)
)
}
```

You can see that this looks very similar to Java code. Let's see what happens here in more detail, as it might look complicated:

1. The MySimulation class extends the Gatling Simulation class. This is the base class for running performance tests.

2. We use the karateProtocol() function to define an instance of the Karate performance testing protocol. This enables Karate to handle the API calls for Gatling and lets Gatling take care of the rest.

3. We set the nameResolver property of the protocol, which accepts two parameters:

    *   req is the HTTP request that is used during the test

    *   ctx is the context, which includes information about the executed test scenario

    The nameResolver function is used to resolve the name of the request that is being sent. In this case, it extracts the value of the karate-name header from the request.

    The protocol.runner.karateEnv("perf") line sets the karateEnv property of the runner object in the protocol instance to perf. This sets the Karate environment to perf (we talked about Karate environments before in *Chapter 4*, *Running Against Different Environments*).

4. This next code block creates the new Gatling scenario variable, myScenario, with the name My great Scenario. Its exec method tells Gatling to execute the karateFeature method that can run Karate feature files. This method, in turn, points to our performance. feature file, which should be used here:

    ```
    val myScenario = scenario("My great Scenario")
        .exec(karateFeature(       "classpath:examples/
    performance/performance.feature")
          )
    ```

5. The setUp code finally defines the parameter of the simulation and adds it to the myScenario variable. In this case, we want to spawn 10 users during a time span of 10 seconds. Note also that we add the protocol we defined before:

    ```
    setUp(
        myScenario.inject(rampUsers(10).during(10))
    ```

```
        .protocols(protocol)
    )
```

This is all the work we need to do on the Gatling side!

> **More Gatling options**
>
> There are many more options to create virtual users besides the `rampUsers` method we have
> used here – for example, `constantUsersPerSec`, which keeps the number of users at a
> fixed value per second, instead of gradually spawning more or fewer users over the course of
> a time frame.
>
> More information about the various options can be seen in the Gatling documentation
> at `https://gatling.io/docs/gatling/reference/current/core/`
> `injection/#open-mode`.

With this setup finally completed, we can run our performance test.

Running the simulation

We are now ready to reuse our Karate scenario as a performance test. The setup might seem very
complex. However, this only needs to be done once.

Since we have everything set up with an additional Maven profile, all it takes to run the test is
the following:

```
mvn clean test -Pperformance
```

The `-P` Maven option activates our `performance` profile, and the Gatling workflow kicks off.

At the end of the simulation run, we should be able to see its statistics in the logs:

```
---- Global Information --------------------------------------
> request count                          10 (OK=10      KO=0     )
> min response time                      77 (OK=77      KO=-     )
> max response time                      94 (OK=94      KO=-     )
> mean response time                     81 (OK=81      KO=-     )
> std deviation                           5 (OK=5       KO=-     )
> response time 50th percentile          80 (OK=80      KO=-     )
> response time 75th percentile          82 (OK=82      KO=-     )
> response time 95th percentile          90 (OK=90      KO=-     )
> response time 99th percentile          93 (OK=93      KO=-     )
> mean requests/sec                       1 (OK=1       KO=-     )
```

```
---- Response Time Distribution --------------------------
> t < 800 ms                        10 (100%)
> 800 ms < t < 1200 ms               0 (  0%)
> t > 1200 ms                        0 (  0%)
> failed                             0 (  0%)

============================================================
```

Here, we can see things such as how many requests were made in total and how many of these were successful (marked with OK) and unsuccessful (marked with KO).

Additional information includes the minimum, maximum, and mean response time of the API under test as well as the distribution of response times. Everything under 800 milliseconds is considered good while everything above 1200 milliseconds is considered bad.

In this case, we can see that everything succeeded, and the API's response times were quite good overall.

Using tags

In our example, we run all scenarios in a specific feature. Karate Gatling also allows you to fine-tune the scenario selection by tags so that you can explicitly choose which scenarios to run as performance tests.

The easiest way to do this is by adding a tag name directly to the feature path:

```
.exec(karateFeature(
  "classpath:examples/performance/performance.feature@perf"
))
```

Here, we run all scenarios within the performance.feature file that are tagged with the @ perf tag.

Typically, that is all you need. However, if you have more complex use cases, please look at the official documentation for more elaborate tagging options: https://karatelabs.github.io/ karate/karate-gatling/#ignore-tags.

Let's change our simulation parameters now to find out more about when our API under testing starts behaving out of the ordinary.

Testing different simulations

It is always a good idea in performance testing to test different simulations to accurately determine the circumstances in which requests fail or their response times take too long.

In this example, I will change the number of users to 1000 for 10 seconds to see the API behavior in this more extreme case:

```
setUp(
  myScenario.inject(rampUsers(1000).during(10))
    .protocols(protocol)
)
```

In this example, we will change the number of users to 1000 for 10 seconds to see the API behavior in this more extreme case.

> **Runtime considerations**
>
> Note that this simulation might run longer than expected. This depends on your system resources. If it is not capable of generating this number of users during the specified time, it will increase the runtime! So, it is a good idea to slowly ramp this up or directly run it on appropriate hardware for large parameters.

Running the simulation with the new parameters again gives us this final output, which does not look as good as before:

```
---- Global Information ----------------------
> request count                       1000 (OK=48        KO=952    )
> min response time                     71 (OK=73        KO=71     )
> max response time                    133 (OK=133       KO=101    )
> mean response time                    77 (OK=86        KO=77     )
> std deviation                          6 (OK=21        KO=3      )
> response time 50th percentile         77 (OK=77        KO=77     )
> response time 75th percentile         79 (OK=80        KO=79     )
> response time 95th percentile         81 (OK=133       KO=81     )
> response time 99th percentile         98 (OK=133       KO=88     )
> mean requests/sec                 90.909 (OK=4.364    KO=86.545)
---- Response Time Distribution ----
> t < 800 ms                            48 (  5%)
> 800 ms < t < 1200 ms                   0 (  0%)
> t > 1200 ms                            0 (  0%)
> failed                               952 ( 95%)
---- Errors -----------------------------------------------------
> examples/performance/performance.feature:7
  status 200                    952 (100,0%)
```

It is clearly visible that 952 of our 1000 requests failed. The last line shows that 100% of these failures had a wrong status code. It even shows exactly in which line of the feature file these errors occurred – another example of the tight integration of Karate and Gatling!

It seems like the API started misbehaving after too many requests. In the logs prior to the result, we can see messages like this multiple times, which strengthens our suspicion:

```
status code was: 429, expected: 200, response time in
milliseconds: 76, url: https://api.jikan.moe/v4/anime/1,
response:
{"status": "429", "type": "RateLimitException", "message": "You
are being rate-limited. Please follow Rate Limiting guidelines:
https://docs.api.jikan.moe/#section/Information/Rate-Limiting",
"error": null}
```

The *Jikan API* documentation confirms this behavior under the *Rate Limiting* section (https://docs.api.jikan.moe/#section/Information/Rate-Limiting):

Rate Limiting

Duration	Requests
Monthly	**Unlimited**
Per Minute	60 requests
Per Second	3 requests

Figure 10.4 – Jikan API rate limiting information

We can see that this API is protected by a **rate limit**, which restricts the number of requests to *60 per minute* and *3 per second*. For this reason, the first requests are successful, and later in the test execution, we receive error responses because the API rejects our flood of requests.

More on rate limiting and throttling

Rate limiting and throttling are techniques used to control the amount and frequency of incoming traffic to a server or network.

Rate limiting sets a maximum number of requests or messages that can be sent within a certain time frame. Typically, you will get a return code of 429 Too Many Requests if this happens (as seen in the Jikan API example with 1,000 users).

Throttling limits the speed at which requests or messages are sent but does not limit the number of requests.

Both techniques are used to prevent server overload, improve performance, and enhance security by blocking or delaying excessive or malicious traffic.

Let's look at a Gatling-specific way to use assertions before moving on to the last section of this chapter.

Gatling assertions

An interesting addition to Karate assertions are assertions directly in the Gatling simulations. These can be added directly to the setup code of a simulation. This has the advantage that performance testing-related assertions such as response times are independent of the Karate tests when run as normal unit tests. They are only active when Gatling is used as the test runner.

For example, this `setUp` block adds two of these assertions:

```
setUp(
    myScenario.inject(rampUsers(10).during(10))
        .protocols(protocol)
).assertions(
    forAll.responseTime.max.lt(100),
    forAll.failedRequests.count.is(0)
)
```

Assertions are activated by adding an `assertions` block to the setup and specifying a list of conditions. In this case, we assert that, for all made requests (`forAll`), the maximum response time (`responseTime.max`) should be less than 100 milliseconds (`lt(100)`). Likewise, we can specify that the number of failed requests (`failedRequests.count`) should be exactly zero (`is(0)`).

The official Gatling documentation has a detailed list of what is possible in this regard: `https://gatling.io/docs/gatling/reference/current/core/assertions`.

As you have seen, running Gatling simulations can already tell us a lot with the information in the logs. However, Gatling also creates very informative reports that reveal even more information, which can be vital to optimize the applications or APIs under test. We will look at these next.

Checking out Gatling reports

Gatling creates its own test reports, which are much better suited to its purpose than standard Karate reports. In this section, we will look at what is inside a report and what conclusions we can draw from it.

We can open the report either from the report link that is shown in the logs, or by clicking on the **KARATE RUNNER** tab in VS Code and selecting the `index.html` link under our performance project:

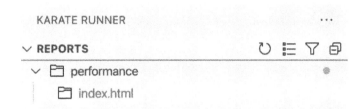

Figure 10.5 – Opening the Gatling report from VS Code

Let's look at the most important individual report sections from the last test run using 1000 users in 10 seconds. This should give us a better understanding of the information we can retrieve from Gatling performance tests.

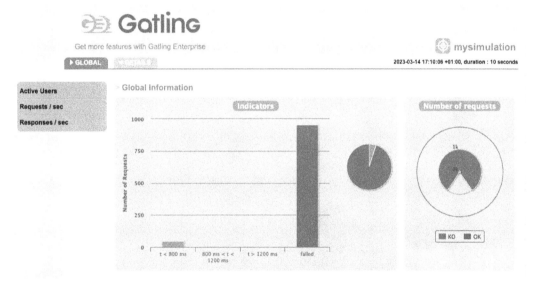

Figure 10.6 – A Gatling report's global simulation data

In *Figure 10.6*, we can see the distribution of response times and passed and failed requests in a much clearer way than in the logs alone. Like in the logs, there are the **KO** and **OK** percentages of requests, as well as the distribution of fast and slow response times. This chart is for the complete test suite, and you cannot see the individually hit endpoints.

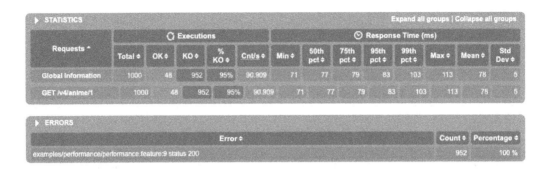

Figure 10.7 – A Gatling report – statistics and error counts

This table in *Figure 10.7* is basically a graphically enhanced version of the last log output section. It shows the total number of executions, the response time percentages, and the counts and percentages of specific errors that occurred. The nice thing about this table is that you can directly see these numbers for the global test run as well as each API that was tested.

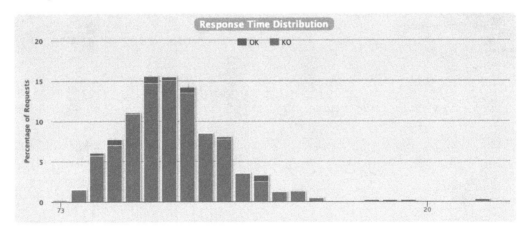

Figure 10.8 – A Gatling report – a response time overview

The chart in *Figure 10.8* shows the distribution of response times. That means how often a specific response time occurred during the test run. Each bar represents a specific response time, and the height of the bars indicates how often this time was hit.

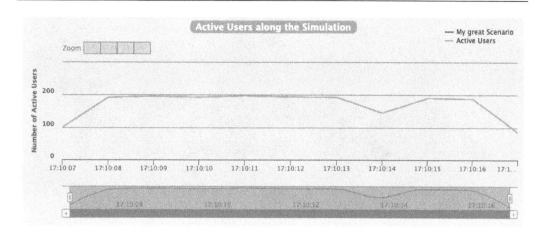

Figure 10.9 – A Gatling report – active user distribution

The chart we can see in *Figure 10.9* shows the number of spawned users over time. In this case, they were quite evenly distributed, as requested. On slower hardware, this curve will be much more uneven on a high number of users.

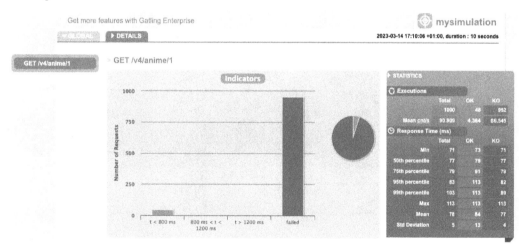

Figure 10.10 – A Gatling report – the DETAILS view

The **DETAILS** view in *Figure 10.10* shows the **OK** and **KO** graphs, as well as the response time distribution, like in the GLOBAL view. However, here you can see it for each individual endpoint. Since we only tested requests to the anime/1 path, this is the only one shown here.

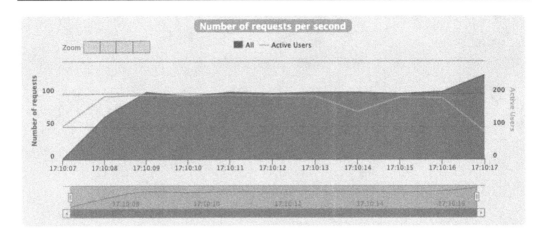

Figure 10.11 – A Gatling report – number of requests per second

The chart shown in *Figure 10.11* can be interesting when judging the effectiveness of the simulation. The solid area shows the number of requests that were made every second, whereas the line on top shows the number of active users that were ramped up.

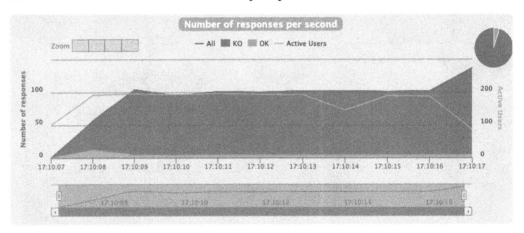

Figure 10.12 – A Gatling report – number of responses per second

Figure 10.12 can be directly compared to the number of requests from *Figure 10.11* previously. It shows the number of responses per second in relation to the ramped-up users.

Also, this chart splits the solid area into two parts, the successful and failed responses.

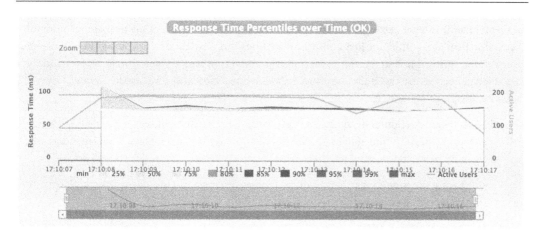

Figure 10.13 – A Gatling report – response time percentiles over time (OK)

This final chart shown in *Figure 10.13* can be tricky to read. It shows the percentage distribution of successful response times over the course of the complete test suite runtime.

Here, you can spot whether responses generally take longer after a certain time frame with a certain number of users, which can indicate problems that gradually occur with longer runtimes under heavy load.

Summary

In this chapter, we looked at how to write performance tests using Karate and Gatling. We covered how to set up a Maven project using profiles so that we can use Karate tests together with Gatling simulations. Finally, we looked at the logs and reports that are generated in these test runs and what insights they can give us.

I hope this chapter has made clear what the combination of Gatling and Karate can do for us. Karate tests typically test the functionality of APIs, whereas Gatling covers the so-called non-functional test cases that are important for the user experience and stability, such as response times, resilience, and data transfer. It can also tell us how much load an application can handle and where performance adjustments need to be made in a system.

This concludes our journey through the world of Karate tests. I hope you were able to learn a few things and apply what you learned to your own projects. As I said before, it is impossible to cover all facets of testing in a book like this. However, I will be happy if what I have written here motivates you to continue your research.

Please remember that when testing software, the most important thing is that the tests are meaningful and cover critical areas. It is not just about finding bugs but also about improving the quality of software and ensuring that it meets the needs and expectations of users. The technology is secondary here, but it can help ensure that a testing project remains successful and sustainable. Karate is a great solution for this.

As the field of software development continues to evolve and new technologies emerge, the importance of software testing will only continue to grow. I encourage you to stay curious, keep learning, and always strive to improve your testing practices.

Thank you for taking the time to read this book and for your commitment to delivering high-quality software. May your testing endeavors be successful, fulfilling, and rewarding.

Index

Packtpub.com

Subscribe to our online digital library for full access to over 7,000 books and videos, as well as industry leading tools to help you plan your personal development and advance your career. For more information, please visit our website.

Why subscribe?

- Spend less time learning and more time coding with practical eBooks and Videos from over 4,000 industry professionals

- Improve your learning with Skill Plans built especially for you

- Get a free eBook or video every month

- Fully searchable for easy access to vital information

- Copy and paste, print, and bookmark content

Did you know that Packt offers eBook versions of every book published, with PDF and ePub files available? You can upgrade to the eBook version at Packtpub.com and as a print book customer, you are entitled to a discount on the eBook copy. Get in touch with us at customercare@packtpub.com for more details.

At www.packtpub.com, you can also read a collection of free technical articles, sign up for a range of free newsletters, and receive exclusive discounts and offers on Packt books and eBooks.

Other Books You May Enjoy

If you enjoyed this book, you may be interested in these other books by Packt:

API Testing and Development with Postman

Dave Westerveld

ISBN: 9781800569201

- Find out what is involved in effective API testing
- Use data-driven testing in Postman to create scalable API tests
- Understand what a well-designed API looks like
- Become well-versed with API terminology, including the different types of APIs
- Get to grips with performing functional and non-functional testing of an API
- Discover how to use industry standards such as OpenAPI and mocking in Postman

Python API Development Fundamentals

Jack Chan, Ray Chung, Jack Huang

ISBN: 9781838983994

- Understand the concept of a RESTful API
- Build a RESTful API using Flask and the Flask-Restful extension
- Manipulate a database using Flask-SQLAlchemy and Flask-Migrate
- Send out plaintext and HTML format emails using the Mailgun API
- Implement a pagination function using Flask-SQLAlchemy
- Use caching to improve API performance and efficiently obtain the latest information
- Deploy an application to Heroku and test it using Postman

Packt is searching for authors like you

If you're interested in becoming an author for Packt, please visit authors.packtpub.com and apply today. We have worked with thousands of developers and tech professionals, just like you, to help them share their insight with the global tech community. You can make a general application, apply for a specific hot topic that we are recruiting an author for, or submit your own idea.

Share Your Thoughts

Now you've finished *Writing API Tests with Karate*, we'd love to hear your thoughts! Scan the QR code below to go straight to the Amazon review page for this book and share your feedback or leave a review on the site that you purchased it from.

https://packt.link/r/1837638268

Your review is important to us and the tech community and will help us make sure we're delivering excellent quality content.

Download a free PDF copy of this book

Thanks for purchasing this book!

Do you like to read on the go but are unable to carry your print books everywhere? Is your eBook purchase not compatible with the device of your choice?

Don't worry, now with every Packt book you get a DRM-free PDF version of that book at no cost.

Read anywhere, any place, on any device. Search, copy, and paste code from your favorite technical books directly into your application.

The perks don't stop there, you can get exclusive access to discounts, newsletters, and great free content in your inbox daily

Follow these simple steps to get the benefits:

1. Scan the QR code or visit the link below

https://packt.link/free-ebook/9781837638260

2. Submit your proof of purchase
3. That's it! We'll send your free PDF and other benefits to your email directly

www.ingramcontent.com/pod-product-compliance
Lightning Source LLC
Chambersburg PA
CBHW080623060326
40690CB00021B/4795